Nonlinear Elasticity

Michel Destrade • Giuseppe Zurlo

Nonlinear Elasticity

A Concise Masterclass for Undergraduates

Michel Destrade
School of Mathematical and Statistical Sciences
University of Galway
Galway, Ireland

Giuseppe Zurlo
School of Mathematical and Statistical Sciences
University of Galway
Galway, Ireland

ISBN 978-3-031-91273-3   ISBN 978-3-031-91274-0 (eBook)
https://doi.org/10.1007/978-3-031-91274-0

© The Editor(s) (if applicable) and The Author(s), under exclusive license to Springer Nature Switzerland AG 2025

This work is subject to copyright. All rights are solely and exclusively licensed by the Publisher, whether the whole or part of the material is concerned, specifically the rights of translation, reprinting, reuse of illustrations, recitation, broadcasting, reproduction on microfilms or in any other physical way, and transmission or information storage and retrieval, electronic adaptation, computer software, or by similar or dissimilar methodology now known or hereafter developed.
The use of general descriptive names, registered names, trademarks, service marks, etc. in this publication does not imply, even in the absence of a specific statement, that such names are exempt from the relevant protective laws and regulations and therefore free for general use.
The publisher, the authors and the editors are safe to assume that the advice and information in this book are believed to be true and accurate at the date of publication. Neither the publisher nor the authors or the editors give a warranty, expressed or implied, with respect to the material contained herein or for any errors or omissions that may have been made. The publisher remains neutral with regard to jurisdictional claims in published maps and institutional affiliations.

This Springer imprint is published by the registered company Springer Nature Switzerland AG
The registered company address is: Gewerbestrasse 11, 6330 Cham, Switzerland

If disposing of this product, please recycle the paper.

To our friends and families.

Preface

The topic of Nonlinear Elasticity is foundational to our understanding of the behaviour of soft matter under various forces. Yet, it remains a challenging area for many undergraduate students because it requires a high level of abstraction and advanced mathematical tools. This monograph, *Nonlinear Elasticity: A Concise Masterclass for Undergraduates*, aims to address this challenge by providing a clear, concise, and accessible introduction to this complex subject. Our motivation for writing this book stemmed from observing that the existing literature often assumes a high level of prior knowledge and mathematical sophistication, which can be daunting for those new to the field.

The origin of this work lies in our combined experiences teaching nonlinear elasticity to undergraduate and graduate students alike, at the University of Galway and beyond (Bari, Brescia, Hangzhou, Limerick, Rome, and online). We observed that students are often enthusiastic about the practical applications of elasticity in engineering and biological contexts, and we aimed to demystify nonlinear elasticity through a structured, methodical, and simple approach.

Our purpose in writing this book is twofold. First, we aim to provide the readers with a solid foundation in nonlinear elasticity, preparing them for advanced studies and professional applications. Second, we seek to make the subject more approachable by presenting it in a clear and engaging manner. By focusing on the elastic response of common materials such as rubber and biological tissues, we illustrate the relevance of nonlinear elasticity in both traditional engineering and emerging biomedical fields.

This short monograph covers the essential principles and mathematical tools necessary for a thorough understanding of nonlinear elasticity, from the basic concepts of stress and strain to more advanced topics such as deformation analysis and constitutive equations. The mathematical prerequisites are minimal, and include basic linear algebra and calculus; they are in any case recalled at the beginning of the book. Each chapter builds on the previous one, to ensure a coherent and logical progression of ideas. We have also included numerous examples and solved some problems to reinforce the theoretical material and provide practical insights.

The plan of the book is organised to facilitate learning. We begin with an introductory chapter that provides historical and conceptual context for the theory of nonlinear elasticity, followed by essential algebra and calculus concepts. The core of the book is dedicated to exploring deformations, stress, and constitutive equations in detail. We conclude with a chapter on solved problems, which serves as a practical application of the theories discussed and opens up new avenues, such as multi-physics coupling.

Our intended audience includes undergraduate students in applied mathematics, physics, and engineering, as well as professionals seeking a concise yet comprehensive reference on nonlinear elasticity. Writing this monograph has been a collaborative and iterative process, and we are deeply grateful to the many colleagues and students who contributed their time, insights, and feedback. Their support and encouragement have been instrumental in shaping this book into a cohesive and effective educational tool. We also acknowledge the contributions of the wider academic community, whose research and publications have provided the foundation for much of the material presented here.

To the reader interested in dwelling deeper into the topic, we recommend the books by Atkin and Fox [1], Chadwick [2], Ericksen [3], Fu and Ogden [4], Gurtin [5], Holzapfel [6], Lai et al. [7], Ogden [8, 9], and Volokh [10]. Accompanying video lectures of our course, recorded during the pandemic, can easily be found online.

Galway, Ireland Michel Destrade
January 2025 Giuseppe Zurlo

References

1. R.J. Atkin, N. Fox, *An Introduction to the Theory of Elasticity*, Dover, New York NY (2005).
2. P. Chadwick, *Continuum Mechanics*, Dover, New York NY (1999).
3. J.L. Ericksen, *Introduction to the Thermodynamics of Solids*, Springer, New York NY (1998).
4. Y.B. Fu, R.W. Ogden (Eds.), *Nonlinear Elasticity: Theory and Applications*, University Press, Cambridge (2001).
5. M.E. Gurtin, *An Introduction to Continuum Mechanics*, Academic Press, Cambridge MA (1981).
6. G.A. Holzapfel, *Nonlinear Solid Mechanics*, Wiley, Hoboken NJ (2000).
7. W.M. Lai, D. Rubin, E. Krempl, *Introduction to Continuum Mechanics, 4th Ed*, Elsevier, Amsterdam (2009).
8. R.W. Ogden, *Non-Linear Elastic Deformations*, Dover, New York NY (1997).

9. R.W. Ogden, *Nonlinear Elasticity with Application to Material Modelling, Lecture Notes*, Centre of Excellence for Advanced Materials and Structures, Institute of Fundamental Technological Research, Polish Academy of Sciences Press, Warsaw (2003).
10. K. Volokh, *Mechanics of Soft Materials*, Springer, New York NY (2016).

Declarations

Competing Interests The authors have no conflicts of interest to declare that are relevant to the content of this book.

Contents

1 **Introduction** 1
2 **Deformations** 17
3 **Stress** 39
4 **Constitutive Equations** 51
5 **Some Solved Problems of Nonlinear Elasticity** 65

Glossary 103
Index 105

Chapter 1
Introduction

1.1 Context

In its beginnings, nonlinear elasticity was seen as an extension of linear elasticity to include "quadratic" terms. With the pioneering and unifying vision of Ronald Rivlin [1915–2005], it came of its own as a discipline in the 1950s to include finite (i.e. large) deformations, with no restrictions on their magnitude. Rivlin sought exact solutions to the general equations of motion of an elastic continuum, instead of performing successive approximations. He was primarily motivated by the desire to model the mechanical behaviour of *rubber* when subjected to large extensions. Since the 1980s, and even more so since the 2000s, an intense effort has been taking place worldwide to apply, transpose, and extend the theories of nonlinear elasticity to the modelling of *biological soft tissues* such as skin, arterial wall, cardiac muscle, brain, tumours, and tendons. Its ultimate goal is to provide accurate computer simulations for applications in mechanical engineering, bioengineering, and bio-medicine.

This book aims at showing how the equations of Continuum Mechanics can be used to provide a sound framework for the description, interpretation, and prediction of the mechanical behaviour of rubber-like materials and soft tissues, by focusing on their *elastic* response.

Elasticity is a concept that is well understood intuitively. Take a highly deformable body, such as a rubber band. Apply a tensile force: it stretches; remove the force: it returns to its original size and shape instantly.

From the points of view of Experimental Physics, Chemistry, Engineering, and Biology, it must be kept in mind that elasticity is an *idealisation*. Nonetheless, nonlinear elasticity has been, and continues to be, very useful for many real-world applications including the design and modelling of aircraft structures, bridges, tyres, mountings, bearings, earthquake protection devices, etc., and of several biomaterials and real tissues used in bioengineering and bio-medicine. It also provides most useful tools in the evaluation, prospection, and visualisation of the Earth's crust and of mechanical and biological structures.

From the points of view of Theoretical Physics and Applied Mathematics, the theory of nonlinear elasticity relies on a wide variety of skills and topics, from linear and tensor algebra, to statistical and data analysis, to calculus and differential geometry. In this introductory book, we only rely on the typical undergraduate mathematics toolkit.

Linear elasticity is concerned with the mechanics of solids for which stress is proportional to the elongation when a slender strip of material is put under a tensile force, say. Stress, strain, stretch, and elongation are notions to be explained in due course, but for the time being, we understand them intuitively in the context of a tensile test. Hence the stress σ is a measure of the amount of force it takes to deform the strip divided by its cross-sectional area. Strain is a measure of the amount of deformation, typically tracked by the stretch $\lambda = \ell/L$, which is the ratio of the current length ℓ of the strip divided by its initial length L, or by the elongation $e = \lambda - 1 = (\ell - L)/L$, which measures the relative change in length.

Figure 1.1a shows a typical response curve for a metal. There, the point P represents the limit of proportionality and elasticity. Beyond P, the body deforms plastically (it does not return to its original shape and state once the load is removed) and eventually breaks. Experimentally, measurements on metals show that P corresponds to tiny strains, typically in the range 0.01%–0.1% for e. Because e is so small, all the equations may be linearised: we retain only linear terms in the Taylor series expansions of the equations with respect to e and neglect all the terms involving higher orders (e^2, e^3, etc.) Then the governing equations are linear, and there is a wealth of tools available to solve linear problems. This is the realm of the so-called infinitesimal, or linear, or *classical*, theory of elasticity.

Figure 1.1b shows a typical tensile response for a biological soft tissue. Here, the point P corresponds to elongations of about 5–10%, and the tissue may be deformed much further, up to a maximum in the range of 20–100% (think, for example, of

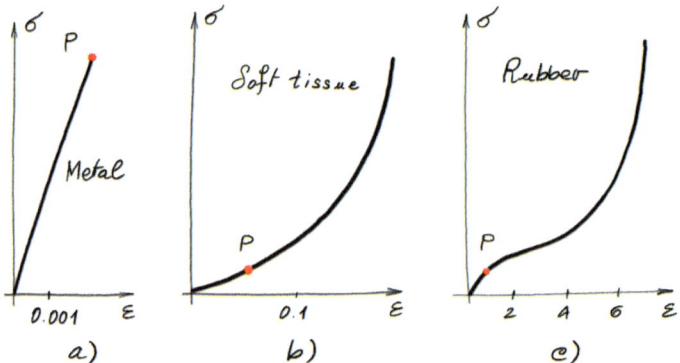

Fig. 1.1 Typical stress-strain response curves in tensile tests of (**a**) metal, (**b**) soft tissue, and (**c**) rubber

1.1 Context

how much a membrane such as the bladder or the skin can stretch). The part of the curve beyond P corresponds to the "strain-stiffening" effect. This J-shape response is often modelled with an exponential type of function. Here linearisation in terms of e is not appropriate, because the strains are *finite*.

Figure 1.1c shows a typical tensile response for rubber-like materials. Rubbers, silicones, and elastomers can stretch up to 100%–1500% ($\lambda = 2$–16). Typically, two changes of curvature occur along this N-shaped curve, with a strong stiffening effect before rupture. Clearly here, linear elasticity can only cover the very early part of the curve.

The initial slope of the tensile test plots gives a measure of the initial stiffness of a given solid. For example, the slope for steel is at least 1000 times steeper than the initial slope for rubber, and 100 times steeper than the initial slope for tendon tissue (which is one of the stiffest biological soft tissues, made of more than 80% collagen in mass). For this reason, rubbers and tissues are often called "soft" solids.

These different behaviours can be interpreted in terms of different *microscopic* structures. Hence, a typical metal has an atomic lattice structure, with movements governed by short-range strong forces, which allow only infinitesimal strains. Biological soft tissues are essentially made of a soft "matrix" with embedded stiffer collagen fibres. Typically, collagen fibres are three orders of magnitude stiffer than the material making up the matrix, and they are crumpled in an undeformed tissue. Hence, the tissue is easily deformed at first, as the fibres unfold, and only the soft elastin contributes to the response. Eventually, the fibres become taut, and their contribution is mostly felt in the late, strain-stiffening part of the curve. A typical rubber-like material is made up of long chain molecules, which are spread randomly and can move quite loosely one with respect to another, with a few points of contact. This setup allows for large strains, until all the chains are aligned with the main direction of tension and they then stiffen the response. See Fig. 1.2 for a schematisation of these microscopic structures.

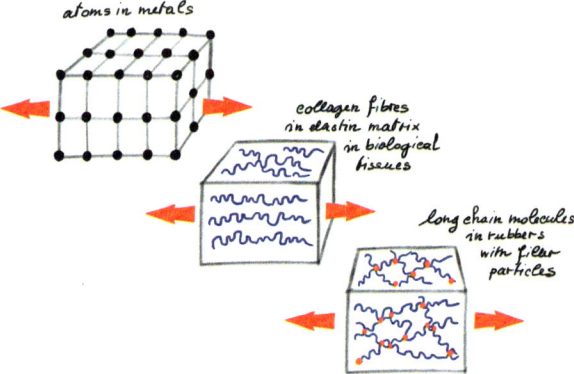

Fig. 1.2 Sketches of micro-structure for (**a**) metal, (**b**) rubber, and (**c**) soft tissue

1.2 Tabletop Nonlinear Elasticity

Stress-strain curves such as those presented in Fig. 1.1 are routinely obtained in mechanical testing facilities, typically using what is called a *tensile machine*, which records both quantities as they change during an automated test. However, it is a simple matter to obtain quite accurately a curve such as Fig. 1.1c with a "tabletop experiment".

For example, we can attach a large empty water bottle to a rubber band, and fill it up progressively in steps, using a full small bottle of water as a loading unit. Then we take a photograph at each step, using a camera fixed at a given spot. The number n of refills is directly proportional to the stress exerted on the band (in a manner that we explicit later in the book). The stretch λ is computed as the ratio of the current length of the rubber band divided by its length when the large bottle was empty (here we neglect that bottle's weight). That ratio is easy to obtain from the photographs, and it is a measure of strain.

In that simple way, we obtain a smooth, nonlinear stress-stretch curve, such as the one displayed in Fig. 1.3. That N-shaped curve is typical of rubber-like materials, such as natural rubber, vulcanised rubber, polyurethane, and silicone. The figure also shows that rubber can easily sustain more than 500% extension.

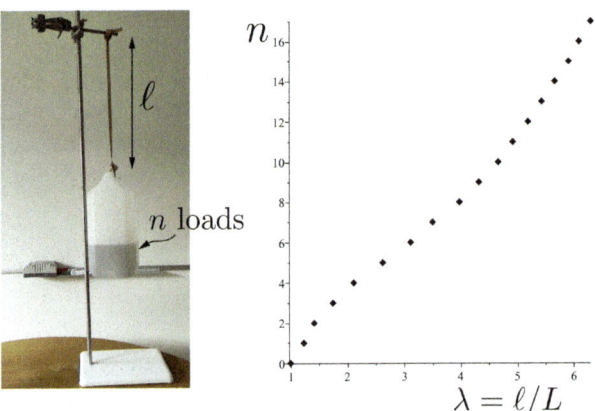

Fig. 1.3 Nonlinear stress-stretch curve readily obtained from a tabletop experiment. The big bottle was loaded by $n = 0, 1, \ldots, 17$ equal refills until the band snapped. The stretch was computed by dividing the current length ℓ of the band by its initial length L, simply by comparing lengths on the digital pictures

1.3 Some Elements of Algebra

Nonlinear elasticity brings together a wide spectrum of mathematical tools. In this book, we rely on algebraic and differential calculus, and linear algebra, for which we need to recall some notions and introduce others that might be new to the readers.

In Physics, we start with *scalar quantities*: time, mass, temperature, voltage, etc., for which we only need to know the magnitude (in a given system of units). Later, we realise that some physical quantities cannot be described by scalars alone, as not only their magnitude is needed but also their direction. And so we introduce the concept of *vector quantities* for velocities, accelerations, forces, magnetic and electric fields, etc. These can be represented by an array of three numbers, in a given coordinate system (in a three-dimensional representation of the physical world).

In this book, we come across physical quantities that cannot be described by scalars or vectors alone. We have to introduce the concept of *tensor quantities*, and we represent them using *matrices*. Hence, a tensor is an algebraic object that is "one step up" from vectors, just as vectors are one step up from scalars. A scalar is represented by one number (say ϕ). In the three-dimensional space as described by a given coordinate system with an origin and an orthonormal basis, a vector (say **u**) is represented by an array of three numbers (its components u_1, u_2, and u_3), and a tensor (say **A**) is represented by a double array of nine numbers (the matrix [A], with components A_{ij}, where $i, j = 1, 2, 3$). Strictly speaking, scalars, vectors, and tensors are all "tensors": scalars are tensors of rank 0, vectors are tensors of rank 1, and what we call "tensors", in this course, are tensors of rank 2. And of course, there are tensors of rank 3, 4, etc., but we shall not need them in this course. As a shorthand, from now on, we use the word "tensor" to designate tensors of rank 2.

Here are some quick reminders and (perhaps) some new notions of linear algebra.

1.4 Vectors

For a fixed coordinate system with an origin O and three mutually orthogonal unit vectors ($\mathbf{e}_1, \mathbf{e}_2, \mathbf{e}_3$), referred to as a *basis*, we can represent two vectors **u** and **v** as follows:

$$\mathbf{u} = u_1\mathbf{e}_1 + u_2\mathbf{e}_2 + u_3\mathbf{e}_3, \quad \mathbf{v} = v_1\mathbf{e}_1 + v_2\mathbf{e}_2 + v_3\mathbf{e}_3,$$

$$\text{or} \quad [\mathbf{u}] = \begin{bmatrix} u_1 \\ u_2 \\ u_3 \end{bmatrix}, \quad [\mathbf{v}] = \begin{bmatrix} v_1 \\ v_2 \\ v_3 \end{bmatrix}, \quad (1.1)$$

where u_i and v_i are the components of the vectors **u** and **v** in the basis ($\mathbf{e}_1, \mathbf{e}_2, \mathbf{e}_3$), and, from now on, we use the square bracket convention to denote arrays of components.

We then define the *dot product* of the two vectors **u** and **v** as the following scalar quantity:

$$\mathbf{u} \cdot \mathbf{v} = \sum_{i=1}^{3} u_i v_i = u_1 v_1 + u_2 v_2 + u_3 v_3. \tag{1.2}$$

The dot product is useful to calculate the *magnitude* $|\mathbf{u}|$ of a vector **u**, as

$$|\mathbf{u}| = \sqrt{\mathbf{u} \cdot \mathbf{u}} = \sqrt{\sum_{i=1}^{3} u_i u_i}. \tag{1.3}$$

The *cross product* of two vectors **u** and **v** gives the vector $\mathbf{u} \times \mathbf{v}$ with components

$$[\mathbf{u} \times \mathbf{v}] = \begin{bmatrix} u_1 \\ u_2 \\ u_3 \end{bmatrix} \times \begin{bmatrix} v_1 \\ v_2 \\ v_3 \end{bmatrix} = \begin{bmatrix} u_2 v_3 - u_3 v_2 \\ -u_1 v_3 + u_3 v_1 \\ u_1 v_2 - u_2 v_1 \end{bmatrix}, \tag{1.4}$$

in the $(\mathbf{e}_1, \mathbf{e}_2, \mathbf{e}_3)$ basis.

Then recall that

$$\mathbf{u} \cdot \mathbf{v} = |\mathbf{u}|\,|\mathbf{v}|\cos(\mathbf{u},\mathbf{v}), \qquad |\mathbf{u} \times \mathbf{v}| = |\mathbf{u}|\,|\mathbf{v}|\sin(\mathbf{u},\mathbf{v}). \tag{1.5}$$

It follows that $|\mathbf{u} \times \mathbf{v}|$ is the area of the parallelogram with edges **u** and **v**, and that the *triple product* $(\mathbf{u} \times \mathbf{v}) \cdot \mathbf{w}$ is the volume of the parallelepiped with edges **u**, **v**, **w**; see Fig. 1.4. A simple way to compute the triple product is to write it as the following determinant (see definition of a determinant in the next section).

$$(\mathbf{u} \times \mathbf{v}) \cdot \mathbf{w} = \begin{vmatrix} u_1 & u_2 & u_3 \\ v_1 & v_2 & v_3 \\ w_1 & w_2 & w_3 \end{vmatrix}. \tag{1.6}$$

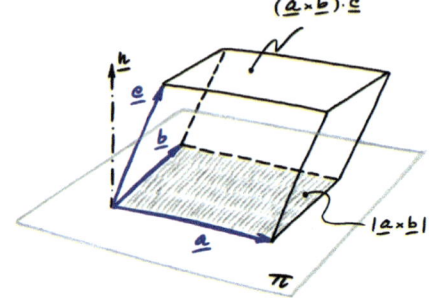

Fig. 1.4 The magnitude of the cross product $\mathbf{a} \times \mathbf{b}$ gives the area of the parallelogram with edges **a** and **b**; the triple product $(\mathbf{a} \times \mathbf{b}) \cdot \mathbf{c}$ gives the area of the parallelepiped with edges **a**, **b**, and **c**

1.5 Tensors

In a given rectangular coordinate system, a tensor **A** is represented by a matrix [A], a square 3 × 3 array of nine elements, the *components* of **A** in that basis:

$$[A] = [A_{ij}] = \begin{bmatrix} A_{11} & A_{12} & A_{13} \\ A_{21} & A_{22} & A_{23} \\ A_{31} & A_{32} & A_{33} \end{bmatrix}. \tag{1.7}$$

The *multiplication of a tensor by a vector gives a vector*. The rule for the components is as follows:

$$[Au] = [A][u] = \begin{bmatrix} A_{11} & A_{12} & A_{13} \\ A_{21} & A_{22} & A_{23} \\ A_{31} & A_{32} & A_{33} \end{bmatrix} \begin{bmatrix} u_1 \\ u_2 \\ u_3 \end{bmatrix} = \begin{bmatrix} A_{11}u_1 + A_{12}u_2 + A_{13}u_3 \\ A_{21}u_1 + A_{22}u_2 + A_{23}u_3 \\ A_{31}u_1 + A_{32}u_2 + A_{33}u_3 \end{bmatrix}. \tag{1.8}$$

Hence, the *i*-th component of the [**Au**] vector is $\sum_{j=1}^{3} A_{ij} u_j$.

Similarly, the *multiplication of a tensor by a tensor gives a tensor*, with the following components:

$$[AB] = \begin{bmatrix} A_{11} & A_{12} & A_{13} \\ A_{21} & A_{22} & A_{23} \\ A_{31} & A_{32} & A_{33} \end{bmatrix} \begin{bmatrix} B_{11} & B_{12} & B_{13} \\ B_{21} & B_{22} & B_{23} \\ B_{31} & B_{32} & B_{33} \end{bmatrix}$$

$$= \begin{bmatrix} A_{11}B_{11} + A_{12}B_{21} + A_{13}B_{31} & A_{11}B_{12} + A_{12}B_{22} + A_{13}B_{32} & A_{11}B_{13} + A_{12}B_{23} + A_{13}B_{33} \\ A_{21}B_{11} + A_{22}B_{21} + A_{23}B_{31} & A_{21}B_{12} + A_{22}B_{22} + A_{23}B_{32} & A_{21}B_{13} + A_{22}B_{23} + A_{23}B_{33} \\ A_{31}B_{11} + A_{32}B_{21} + A_{33}B_{31} & A_{31}B_{12} + A_{32}B_{22} + A_{33}B_{32} & A_{31}B_{13} + A_{32}B_{23} + A_{33}B_{33} \end{bmatrix}. \tag{1.9}$$

Hence, the *ij*-th component of the [AB] matrix is $\sum_{k=1}^{3} A_{ik} B_{kj}$. Note that in general, and in contrast to numbers algebra, tensors do not commute (in general, $AB \neq BA$). The following power notations are also used, $A^2 = AA$, $A^3 = AAA$, and so on.

The matrix of the *transpose* A^T of a tensor **A** has elements flipped over its diagonal, so rows become columns and vice versa:

$$[A^T] = [A_{ij}]^T = [A_{ji}] = \begin{bmatrix} A_{11} & A_{21} & A_{31} \\ A_{12} & A_{22} & A_{32} \\ A_{13} & A_{23} & A_{33} \end{bmatrix}. \tag{1.10}$$

Then it is easy to show that the following identities hold:

$$(AB)^T = B^T A^T, \qquad \mathbf{u} \cdot A\mathbf{v} = \mathbf{v} \cdot A^T \mathbf{u}, \tag{1.11}$$

for any vectors **U**, **v** and tensors **A**, **B**.

A *symmetric tensor* is such that $\mathbf{A} = \mathbf{A}^\mathsf{T}$. Hence, in matrix and component terms, we have $[\mathbf{A}] = [\mathbf{A}^\mathsf{T}]$ and $A_{ij} = A_{ji}$. In other words, the matrix of a symmetric tensor displays a symmetry of the off-diagonal components with respect to its main diagonal and thus has six independent components at most.

An important example of symmetric tensor is the *unit (or identity) tensor*, **I**; its components in any coordinate system are

$$[\mathbf{I}] = [\delta_{ij}] = \begin{bmatrix} 1 & 0 & 0 \\ 0 & 1 & 0 \\ 0 & 0 & 1 \end{bmatrix}, \quad (1.12)$$

where δ_{ij}, the *Kronecker delta*, is such that $\delta_{ij} = 1$ when $i = j$ and $\delta_{ij} = 0$ when $i \neq j$. This gives

$$\sum_{j=1}^{3} \delta_{ij} A_{jk} = \sum_{j=1}^{3} A_{ij} \delta_{jk} = A_{ik}, \quad \mathbf{IA} = \mathbf{AI} = \mathbf{A}, \quad (1.13)$$

so that the product of any tensor **A** with **I** is **A**, which is why **I** is called the identity matrix.

The *trace* of a 3×3 square matrix is the sum of the terms on its leading diagonal:

$$\mathrm{tr}[\mathbf{A}] = A_{11} + A_{22} + A_{33}. \quad (1.14)$$

In particular, note that $\mathrm{tr}[\mathbf{I}] = 3$.

The *determinant* of a 2×2 matrix $[\mathbf{B}]$ is defined as

$$\det[\mathbf{B}] = \begin{vmatrix} B_{11} & B_{12} \\ B_{21} & B_{22} \end{vmatrix} = B_{11} B_{22} - B_{12} B_{21}. \quad (1.15)$$

Then the determinant of a 3×3 matrix $[\mathbf{A}]$ can be calculated from (1.15) as follows:

$$\det[\mathbf{A}] = \begin{vmatrix} A_{11} & A_{12} & A_{13} \\ A_{21} & A_{22} & A_{23} \\ A_{31} & A_{32} & A_{33} \end{vmatrix} = A_{11} \begin{vmatrix} A_{22} & A_{23} \\ A_{32} & A_{33} \end{vmatrix} - A_{12} \begin{vmatrix} A_{21} & A_{23} \\ A_{31} & A_{33} \end{vmatrix} + A_{13} \begin{vmatrix} A_{21} & A_{22} \\ A_{31} & A_{32} \end{vmatrix}. \quad (1.16)$$

With the definition of the determinant, we can prove (but it is a lengthy task) the following important properties:

$$\det(\mathbf{A}^\mathsf{T}) = \det \mathbf{A}, \quad \det(\mathbf{AB}) = (\det \mathbf{A})(\det \mathbf{B}), \quad \det(\mathbf{I}) = 1. \quad (1.17)$$

1.6 Components and Bases

In particular, we see from the first property that we can rewrite the formula for the triple product (1.6) as

$$(\mathbf{u} \times \mathbf{v}) \cdot \mathbf{w} = \begin{vmatrix} u_1 & v_1 & w_1 \\ u_2 & v_2 & w_2 \\ u_3 & v_3 & w_3 \end{vmatrix}, \tag{1.18}$$

which is the determinant of the matrix made by staking the components of $\mathbf{u}, \mathbf{v}, \mathbf{w}$ as columns.

In numbers algebra, we say that x has an inverse x^{-1}, such that $xx^{-1} = 1$, provided $x \neq 0$. The counterpart statement in tensor algebra is that a tensor $\mathbf{A}$ has an *inverse* $\mathbf{A}^{-1}$, such as $\mathbf{A}\mathbf{A}^{-1} = \mathbf{I}$, provided $\det \mathbf{A} \neq 0$. Such a tensor is said to be *non-singular*. Now it is easy to establish that for a non-singular $\mathbf{A}$, we have

$$\det(\mathbf{A}^{-1}) = \frac{1}{\det \mathbf{A}}, \qquad \det(\mathbf{A}^{-1}\mathbf{B}) = \frac{\det \mathbf{B}}{\det \mathbf{A}}. \tag{1.19}$$

An *orthogonal tensor* $\mathbf{Q}$ is such that $\mathbf{Q}^{-1} = \mathbf{Q}^\mathsf{T}$. It then follows, by taking the determinant of $\mathbf{Q}\mathbf{Q}^\mathsf{T} = \mathbf{I}$ and using the relationships above, that $\det \mathbf{Q} = \pm 1$. A *proper orthogonal tensor* $\mathbf{R}$, also known as a *rotation*, is such that

$$\mathbf{R}\mathbf{R}^\mathsf{T} = \mathbf{I}, \qquad \det \mathbf{R} = +1. \tag{1.20}$$

1.6 Components and Bases

Once we adopt a given system of units, say the *International System of Units*, or SI for brief (from French *Système international d'unités*), the value of a scalar does not change from one coordinate system to another.

Similarly, the characteristics of a vector (its direction, sense, and magnitude) do not change when we move from one coordinate system to another. However, its components are different in different orthonormal bases.

For example, consider two bases $(\mathbf{e}_1, \mathbf{e}_2, \mathbf{e}_3)$ and $(\mathbf{e}_1^+, \mathbf{e}_2^+, \mathbf{e}_3^+)$, the latter obtained by rotating the former about $\mathbf{e}_3$ by the angle θ, so that (see Fig. 1.5)

$$\mathbf{e}_1 = (\cos\theta)\mathbf{e}_1^+ - (\sin\theta)\mathbf{e}_2^+, \qquad \mathbf{e}_2 = (\sin\theta)\mathbf{e}_1^+ + (\cos\theta)\mathbf{e}_2^+, \qquad \mathbf{e}_3 = \mathbf{e}_3^+. \tag{1.21}$$

Now consider a vector $\mathbf{u}$, with components $[\mathbf{u}]$ in the basis $(\mathbf{e}_1, \mathbf{e}_2, \mathbf{e}_3)$ and $[\mathbf{u}]^+$ in the basis $(\mathbf{e}_1^+, \mathbf{e}_2^+, \mathbf{e}_3^+)$, so that

$$\mathbf{u} = u_1\mathbf{e}_1 + u_2\mathbf{e}_2 + u_3\mathbf{e}_3 = u_1^+\mathbf{e}_1^+ + u_2^+\mathbf{e}_2^+ + u_3^+\mathbf{e}_3^+. \tag{1.22}$$

Fig. 1.5 The characteristics (direction, sense, and magnitude) of a vector **u** are not affected by a change of orthonormal basis, but its components are different in different bases

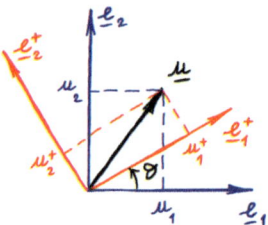

Then using (1.21) into (1.22) reveals that

$$[\mathbf{u}]^+ = \begin{bmatrix} \cos\theta & \sin\theta & 0 \\ -\sin\theta & \cos\theta & 0 \\ 0 & 0 & 1 \end{bmatrix} [\mathbf{u}], \qquad (1.23)$$

where it is clear that the 3×3 matrix is the matrix of a rotation, as it satisfies (1.20). It is easy to show that in general (when the two bases need not have a common unit vector as in this example), the components of a vector change as

$$[\mathbf{u}]^+ = [\mathsf{R}][\mathbf{u}], \qquad (1.24)$$

where R is a rotation.

A similar rule applies to tensors. Consider the vector $\mathsf{T}\mathbf{u}$, where T is a tensor and $\mathbf{u}$ is a vector, and call R the rotation from one basis to the other. Then, according to (1.24), $[\mathsf{T}\mathbf{u}]^+ = [\mathsf{R}\mathsf{T}\mathbf{u}]$. Also, $[\mathsf{T}\mathbf{u}]^+ = [\mathsf{T}]^+[\mathbf{u}]^+$, which gives $[\mathsf{T}]^+[\mathsf{R}][\mathbf{u}] = [\mathsf{R}\mathsf{T}][\mathbf{u}]$ for all $\mathbf{u}$. Then, because R is a rotation, we conclude that

$$[\mathsf{T}]^+ = [\mathsf{R}\mathsf{T}\mathsf{R}^\mathsf{T}] \qquad (1.25)$$

which is the rule to change tensor components from one basis to another.

Just like some features of vectors are independent of the chosen basis, so are some features of tensors, called *invariants*; their role is crucial in nonlinear elasticity, as we start to learn in the next section.

1.7 Eigenvalues and Eigenvectors

For any scalar ω and tensor A, the determinant of $\mathsf{A} - \omega\mathsf{I}$ expands as a cubic in ω,

$$\det(\mathsf{A} - \omega\mathsf{I}) = -\omega^3 + \omega^2 I_1 - \omega I_2 + I_3, \qquad (1.26)$$

1.7 Eigenvalues and Eigenvectors

where I_1, I_2, I_3 are called the first three *principal invariants* of $\mathbf{A}$ and are given by

$$I_1 = \operatorname{tr} \mathbf{A}, \qquad I_2 = \tfrac{1}{2}\left[(\operatorname{tr}\mathbf{A})^2 - \operatorname{tr}\left(\mathbf{A}^2\right)\right], \qquad I_3 = \det \mathbf{A}. \qquad (1.27)$$

That expansion is straightforward (if cumbersome) to check, using (1.16). Note also that the invariants above are written in terms of the components of $\mathbf{A}$ in a given coordinate system, but in fact, thanks to the change of coordinate rule for tensors, they are the same in any coordinate system, hence their name. The proof is simple, because $\operatorname{tr}(\mathbf{R}^\mathsf{T}\mathbf{A}\mathbf{R}) = \operatorname{tr}(\mathbf{R}\mathbf{R}^\mathsf{T}\mathbf{A}) = \operatorname{tr}\mathbf{A}$, and similarly for the determinant.

Now let now $\mathbf{S}$ be a symmetric tensor. The *eigenvalues* of $\mathbf{S}$ are the roots $\lambda_1, \lambda_2, \lambda_3$ of the *characteristic equation* of $\mathbf{S}$

$$\det(\mathbf{S} - \lambda \mathbf{I}) = -\lambda^3 + I_1 \lambda^2 - I_2 \lambda + I_3 = 0, \qquad (1.28)$$

again an equation which is independent of the choice of coordinate system.

Because $\lambda_1, \lambda_2, \lambda_3$ are the roots of that cubic, we also have $-(\lambda - \lambda_1)(\lambda - \lambda_2)(\lambda - \lambda_3) = 0$, so that by expansion and comparison with (1.28), the following identifications apply:

$$I_1 = \lambda_1 + \lambda_2 + \lambda_3, \qquad I_2 = \lambda_1 \lambda_2 + \lambda_2 \lambda_3 + \lambda_3 \lambda_1, \qquad I_3 = \lambda_1 \lambda_2 \lambda_3. \qquad (1.29)$$

By solving the vector equation $\mathbf{Su} = \lambda_1 \mathbf{u}$ (i.e. $(\mathbf{S} - \lambda_1 \mathbf{I})\mathbf{u} = \mathbf{0}$) for $\mathbf{u}$, we find the *eigenvectors* of $\mathbf{S}$ corresponding to λ_1. They are all scalar multiples one of another. We call $\mathbf{r}$ the unit eigenvector ($\mathbf{r} \cdot \mathbf{r} = 1$) corresponding to λ_1. Similarly, we define the unit vectors $\mathbf{s}$ and $\mathbf{t}$ corresponding to λ_2 and λ_3, respectively:

$$\mathbf{Sr} = \lambda_1 \mathbf{r}, \qquad \mathbf{Ss} = \lambda_2 \mathbf{s}, \qquad \mathbf{St} = \lambda_3 \mathbf{t}. \qquad (1.30)$$

Now we prove that *eigenvectors corresponding to distinct eigenvalues are orthogonal*. To do so, first write the following dot product:

$$\mathbf{r} \cdot \mathbf{Ss} = \mathbf{r} \cdot (\lambda_2 \mathbf{s}) = \lambda_2 \mathbf{r} \cdot \mathbf{s}. \qquad (1.31)$$

On the other hand, by (1.11) that dot product is also equal to

$$\mathbf{s} \cdot \mathbf{S}^\mathsf{T} \mathbf{r} = \mathbf{s} \cdot \mathbf{Sr} = \mathbf{s} \cdot (\lambda_1 \mathbf{r}) = \lambda_1 \mathbf{s} \cdot \mathbf{r}. \qquad (1.32)$$

By subtraction, $(\lambda_2 - \lambda_1) \mathbf{r} \cdot \mathbf{s} = 0$, i.e. $\mathbf{r} \cdot \mathbf{s} = 0$.

Similarly, we can prove that the eigenvalues of the real symmetric tensor $\mathbf{S}$ are *real*. Hence, we have

$$\overline{\bar{\mathbf{r}} \cdot \mathbf{Sr}} = \mathbf{r} \cdot \bar{\mathbf{S}}\bar{\mathbf{r}} = \mathbf{r} \cdot \mathbf{S}\bar{\mathbf{r}} = \bar{\mathbf{r}} \cdot \mathbf{S}^\mathsf{T} \mathbf{r} = \bar{\mathbf{r}} \cdot \mathbf{Sr}. \qquad (1.33)$$

Thus, $\bar{\mathbf{r}} \cdot \mathbf{S}\mathbf{r}$ is equal to its complex conjugate and is real. Because it is equal to $\lambda_1 \mathbf{r} \cdot \bar{\mathbf{r}}$, it follows that λ_1 is real. Note that therefore, $\mathbf{r}, \mathbf{s}, \mathbf{t}$ are also real because they are the solutions of an eigensystem of equations with real coefficients (the homogenous system of equations $(\mathbf{S} - \lambda \mathbf{I})\mathbf{u} = \mathbf{0}$).

These results were established by assuming that the cubic (1.28) yields three distinct roots. In the case of one double root plus a single root, or of a triple root, some adjustments have to be made, but the main results remain: we can always find at least one direct orthonormal basis made of real unit eigenvectors of $\mathbf{S}$.

A tensor $\mathbf{S}$ is *positive definite* if $\mathbf{S}\mathbf{v} \cdot \mathbf{v} > 0$ for all vectors $\mathbf{v} \neq \mathbf{0}$. The eigenvalues of a positive-definite tensor $\mathbf{S}$ are strictly positive, because

$$\mathbf{S}\mathbf{r} \cdot \mathbf{r} = \lambda_1 \mathbf{r} \cdot \mathbf{r} = \lambda_1 > 0, \quad (1.34)$$

and likewise for the other eigenvalues.

Finally, we establish an important theorem of tensor algebra, the *Cayley-Hamilton theorem*. Any vector $\mathbf{v}$ can be decomposed in the orthonormal basis $(\mathbf{r}, \mathbf{s}, \mathbf{t})$, as $\mathbf{v} = \alpha \mathbf{r} + \beta \mathbf{s} + \gamma \mathbf{t}$, say, for some scalars α, β, γ. Now consider the following operation:

$$\begin{aligned} & -(\mathbf{S} - \lambda_1 \mathbf{I})(\mathbf{S} - \lambda_2 \mathbf{I})(\mathbf{S} - \lambda_3 \mathbf{I})\mathbf{v} \\ &= -(\mathbf{S} - \lambda_1 \mathbf{I})(\mathbf{S} - \lambda_2 \mathbf{I})[\alpha(\lambda_1 - \lambda_3)\mathbf{r} + \beta(\lambda_2 - \lambda_3)\mathbf{s}] \\ &= -(\mathbf{S} - \lambda_1 \mathbf{I})[\alpha(\lambda_1 - \lambda_3)(\lambda_1 - \lambda_2)\mathbf{r}] = \mathbf{0}, \quad (1.35) \end{aligned}$$

for all $\mathbf{v}$. Hence, $-(\mathbf{S} - \lambda_1 \mathbf{I})(\mathbf{S} - \lambda_2 \mathbf{I})(\mathbf{S} - \lambda_3 \mathbf{I}) = \mathbf{O}$, or, by expanding and using (1.29),

$$-\mathbf{S}^3 + I_1 \mathbf{S}^2 - I_2 \mathbf{S} + I_3 \mathbf{I} = \mathbf{O}. \quad (1.36)$$

In other words, $\mathbf{S}$ *satisfies its own characteristic equation*.

A consequence from the identity above is that $\mathbf{S}^3$ can be expressed in terms of $\mathbf{S}^2$ and $\mathbf{S}$, as $\mathbf{S}^3 = I_1 \mathbf{S}^2 - I_2 \mathbf{S} + I_3 \mathbf{I}$. When we multiply this identity by $\mathbf{S}$, we find that $\mathbf{S}^4$ can be expressed in terms of $\mathbf{S}^3, \mathbf{S}^2$, and $\mathbf{S}$, i.e. in terms of $\mathbf{S}^2$ and $\mathbf{S}$, and similarly for higher powers. In other words, $\mathbf{S}^k$ with $k \geq 3$ can always be written in terms of $\mathbf{S}^2$ and $\mathbf{S}$. Similarly, when we multiply the identity by $\mathbf{S}^{-1}$, we find that $\mathbf{S}^2$ can be expressed in terms of $\mathbf{S}$ and $\mathbf{S}^{-1}$, as

$$\mathbf{S}^2 = I_1 \mathbf{S} - I_2 \mathbf{I} + I_3 \mathbf{S}^{-1}. \quad (1.37)$$

1.8 Jacobi's Identity

Here we present a neat result about the derivative of the determinant of an invertible tensor $\mathbf{A} = \mathbf{A}(t)$ which depends on a scalar parameter t. The *Jacobi identity* reads

$$\frac{d}{dt}(\det \mathbf{A}) = (\det \mathbf{A})\,\mathrm{tr}\left(\mathbf{A}^{-1}\frac{d\mathbf{A}}{dt}\right). \tag{1.38}$$

The proof is quite involved. First, we recall that $\det(a\mathbf{A}) = a^3 \det \mathbf{A}$ in general, so that $\det(\mathbf{M} - \omega \mathbf{I}) = -\omega^{-3}\det(\mathbf{I} - \omega^{-1}\mathbf{M})$, for any tensors $\mathbf{A}, \mathbf{M}$ and constants a, ω. Then, with $\omega = -1/h$, (1.26) leads to

$$\det(\mathbf{I} + h\mathbf{M}) = 1 + hI_1 + h^2 I_2 + h^3 I_3, \tag{1.39}$$

where I_1, I_2, I_3 are the principal invariants of the tensor $\mathbf{M}$. Now we compute the derivative of the determinant from first principles

$$\frac{d}{dt}\det \mathbf{A}(t) = \lim_{h \to 0}\frac{\det \mathbf{A}(t+h) - \det \mathbf{A}(t)}{h}$$

$$= (\det \mathbf{A}(t))\lim_{h \to 0}\frac{\frac{\det \mathbf{A}(t+h)}{\det \mathbf{A}(t)} - 1}{h}$$

$$= (\det \mathbf{A}(t))\lim_{h \to 0}\frac{\det[\mathbf{A}(t)^{-1}\mathbf{A}(t+h)] - 1}{h}, \tag{1.40}$$

using (1.19). But a Taylor expansion gives $\mathbf{A}(t+h) = \mathbf{A}(t) + h\frac{d\mathbf{A}(t)}{dt} + \frac{1}{2}h^2\frac{d^2\mathbf{A}(t)}{dt^2} + \ldots$, so we may set $\mathbf{A}(t)^{-1}\mathbf{A}(t+h) = \mathbf{I} + h\mathbf{M}$, where $\mathbf{M} = \mathbf{A}^{-1}\frac{d\mathbf{A}}{dt} + \frac{1}{2}h\mathbf{A}^{-1}\frac{d^2\mathbf{A}}{dt^2} + \ldots$. Then, it follows from the identity (1.39) that

$$\frac{d}{dt}(\det \mathbf{A}(t)) = (\det \mathbf{A}(t))\lim_{h \to 0}\frac{hI_1 + h^2 I_2 + h^3 I_3}{h} = (\det \mathbf{A}(t))\lim_{h \to 0} I_1, \tag{1.41}$$

where I_k are the principal invariants of $\mathbf{M}$. Noticing that $I_1 = \mathrm{tr}\,\mathbf{M} = \mathrm{tr}(\mathbf{A}^{-1}\frac{d\mathbf{A}}{dt}) + \frac{1}{2}h\,\mathrm{tr}(\mathbf{A}^{-1}\frac{d^2\mathbf{A}}{dt^2}) + \ldots$, and thus that $\lim_{h \to 0} I_1 = \mathrm{tr}(\mathbf{A}^{-1}\frac{d\mathbf{A}}{dt})$, we finally arrive at Jacobi's identity.

1.9 Differential Operators

In this course, we study objects in a three-dimensional world, and (mostly) restrict our attention to rectangular coordinate systems. Hence the scalars, vectors, and tensors are functions of the three coordinates (x_1, x_2, x_3), and they may be differentiated with respect to those coordinates. Here we introduce several ways of differentiating these objects. A mnemonic way of remembering some of the formulas below can be achieved by introducing a "differential vector", the so-called *nabla operator* ∇, with components:

$$[\nabla] = \begin{bmatrix} \dfrac{\partial}{\partial x_1} \\ \dfrac{\partial}{\partial x_2} \\ \dfrac{\partial}{\partial x_3} \end{bmatrix}, \qquad (1.42)$$

in the $(\mathbf{e}_1, \mathbf{e}_2, \mathbf{e}_3)$ basis.

The *gradient* of the scalar $\phi(x_1, x_2, x_3)$ is the vector grad $\phi = \nabla\phi$ with components

$$[\text{grad } \phi] = [\nabla\phi] = \begin{bmatrix} \dfrac{\partial \phi}{\partial x_1} \\ \dfrac{\partial \phi}{\partial x_2} \\ \dfrac{\partial \phi}{\partial x_3} \end{bmatrix}, \qquad [\text{grad } \phi]_i = \dfrac{\partial \phi}{\partial x_i}. \qquad (1.43)$$

The *gradient* of the vector $\mathbf{u}(x_1, x_2, x_3)$ is the tensor grad $\mathbf{u}$, with components

$$[\text{grad } \mathbf{u}] = \begin{bmatrix} \dfrac{\partial u_1}{\partial x_1} & \dfrac{\partial u_1}{\partial x_2} & \dfrac{\partial u_1}{\partial x_3} \\ \dfrac{\partial u_2}{\partial x_1} & \dfrac{\partial u_2}{\partial x_2} & \dfrac{\partial u_2}{\partial x_3} \\ \dfrac{\partial u_3}{\partial x_1} & \dfrac{\partial u_3}{\partial x_2} & \dfrac{\partial u_3}{\partial x_3} \end{bmatrix}, \qquad [\text{grad } \mathbf{u}]_{ij} = \dfrac{\partial u_i}{\partial x_j}. \qquad (1.44)$$

The *divergence* of the vector $\mathbf{u}(x_1, x_2, x_3)$ is the following scalar:

$$\text{div } \mathbf{u} = \nabla \cdot \mathbf{u} = \dfrac{\partial u_1}{\partial x_1} + \dfrac{\partial u_2}{\partial x_2} + \dfrac{\partial u_3}{\partial x_3}, \qquad \text{div } \mathbf{u} = \sum_{i=1}^{3} \dfrac{\partial u_i}{\partial x_i}. \qquad (1.45)$$

1.9 Differential Operators

The *divergence* of the tensor A is the vector div A, with components

$$[\text{div } \mathbf{A}] = \begin{bmatrix} \dfrac{\partial A_{11}}{\partial x_1} + \dfrac{\partial A_{21}}{\partial x_2} + \dfrac{\partial A_{31}}{\partial x_3} \\ \dfrac{\partial A_{12}}{\partial x_1} + \dfrac{\partial A_{22}}{\partial x_2} + \dfrac{\partial A_{32}}{\partial x_3} \\ \dfrac{\partial A_{13}}{\partial x_1} + \dfrac{\partial A_{23}}{\partial x_2} + \dfrac{\partial A_{33}}{\partial x_3} \end{bmatrix}, \qquad [\text{div } \mathbf{A}]_i = \sum_{j=1}^{3} \dfrac{\partial A_{ji}}{\partial x_j}. \qquad (1.46)$$

Note that this is the convention used throughout the book for the divergence operator. Some textbooks do the sum on the second index instead.

Chapter 2
Deformations

When a soft solid is subjected to forces, it deforms: particles that were close to each other are pulled apart or brought closer together. Before we study how to describe the forces that can be applied to a body of soft solid material, we need to be able to describe how its particles are displaced when it is deformed. This chapter provides an analysis of deformation, without reference yet to the forces and moments that produce it.

2.1 Configurations

A soft solid is made of particles so closely packed together that they appear as a continuum to the human eye. The solid is soft so that, under the action of forces, it deforms and the particles, or material points, are displaced with respect to each other.

We look first at the solid when it is at rest, unloaded, in what we call the *reference configuration*, denoted by $\mathcal{B}_r$. Then we examine the soft solid when it is held in a deformed state by the application of forces and moments, in what we call the *current configuration*, denoted by $\mathcal{B}_c$.

Material points are identified by their position vector **X** relative to some origin **O** in the reference configuration $\mathcal{B}_r$, or by the position vector **x** relative to some origin **o** in the current configuration $\mathcal{B}_c$; see Fig. 2.1.

Because $\mathcal{B}_r$ and $\mathcal{B}_c$ are configurations of the same body, and we assume that there is no creation nor annihilation of matter, there exists a one-to-one mapping χ between $\mathcal{B}_r$ and $\mathcal{B}_c$ such that

$$\mathbf{x} = \chi(\mathbf{X}). \tag{2.1}$$

© The Author(s), under exclusive license to Springer Nature Switzerland AG 2025
M. Destrade, G. Zurlo, *Nonlinear Elasticity*,
https://doi.org/10.1007/978-3-031-91274-0_2

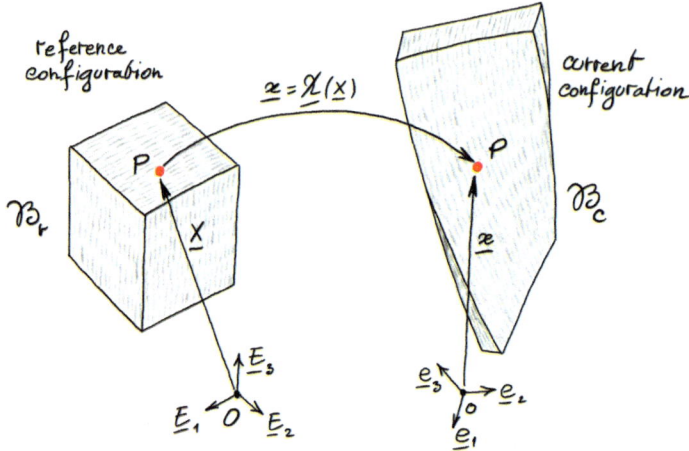

Fig. 2.1 Schematic representation of a soft solid in the reference configuration $\mathcal{B}_r$ and, once deformed through $\mathbf{x} = \chi(\mathbf{X})$, in the current configuration $\mathcal{B}_c$, with the position vectors $\mathbf{X}$ in $\mathcal{B}_r$ and $\mathbf{x}$ in $\mathcal{B}_c$ of a material point P

The mapping χ is called the *deformation* of the body *from $\mathcal{B}_r$ to $\mathcal{B}_c$*, and we assume that the vector function χ is smooth (at least twice-continuously differentiable with respect to position).

In $\mathcal{B}_r$, we use capital letters for quantities, and in $\mathcal{B}_c$, we use lowercase letters. Using rectangular Cartesian coordinate systems with orthonormal vectors $(\mathbf{E}_1, \mathbf{E}_2, \mathbf{E}_3)$ and $(\mathbf{e}_1, \mathbf{e}_2, \mathbf{e}_3)$ in $\mathcal{B}_r$ and $\mathcal{B}_c$, respectively, we have

$$\mathbf{X} = \sum_{i=1}^{3} X_i \mathbf{E}_i, \qquad \mathbf{x} = \sum_{i=1}^{3} x_i \mathbf{e}_i, \tag{2.2}$$

with *material* (or *Lagrangian*) *coordinates* X_i and *spatial* (or *Eulerian*) *coordinates* x_i ($i = 1, 2, 3$). We may express fields, such as density, temperature, strain, and stress, in terms of X_i or x_i. In the first case, we talk of the *referential* or *material* or *Lagrangian* description; in the second case, of the *current* or *spatial* or *Eulerian* description.

Both descriptions are equivalent, but sometimes one is more convenient than the other for a given problem. It is thus important to know how to go from one to the other.

2.2 The Deformation Gradient

Following our convention with typed cases, we let Grad, Div, Curl denote the gradient, divergence, and curl operators, respectively, in the reference configuration $\mathcal{B}_r$, where differentiation is done with respect to $\mathbf{X}$, and we let grad, div, curl denote the gradient, divergence, and curl operators, respectively, in the current configuration $\mathcal{B}_c$, where differentiation is done with respect to $\mathbf{x}$. Hence, for example,

$$[\operatorname{Grad}\phi]_i = \frac{\partial \phi}{\partial X_i}, \qquad [\operatorname{div} \mathsf{T}]_i = \sum_{j=1}^{3} \frac{\partial T_{ji}}{\partial x_j}, \qquad \text{etc.} \qquad (2.3)$$

We then define the *deformation gradient tensor* F as

$$\mathsf{F}(\mathbf{X}) = \operatorname{Grad} \mathbf{x} = \operatorname{Grad} \chi(\mathbf{X}), \qquad (2.4)$$

so that it has components

$$F_{ij} = \frac{\partial x_i}{\partial X_j}. \qquad (2.5)$$

Provided $\det \mathsf{F} \neq 0$ (to be justified shortly), the inverse of F is

$$\mathsf{F}^{-1} = \operatorname{grad} \mathbf{X}, \qquad [\mathsf{F}^{-1}]_{ij} = \frac{\partial X_i}{\partial x_j}. \qquad (2.6)$$

We can check if this is the correct expression for F^{-1} by calculating FF^{-1}, and see that it is indeed the identity tensor I:

$$[\mathsf{FF}^{-1}]_{ij} = \sum_{k=1}^{3} F_{ik} F_{kj}^{-1} = \sum_{k=1}^{3} \frac{\partial x_i}{\partial X_k} \frac{\partial X_k}{\partial x_j} = \frac{\partial x_i}{\partial x_j} = \delta_{ij}, \qquad (2.7)$$

where we used the chain rule.

It follows from (2.5) that

$$\sum_{j=1}^{3} F_{ij} dX_j = \sum_{j=1}^{3} \frac{\partial x_i}{\partial X_j} dX_j = dx_i, \qquad (2.8)$$

or, in vector form,

$$d\mathbf{x} = \mathsf{F} d\mathbf{X}. \qquad (2.9)$$

Inverting that equation, we have

$$d\mathbf{X} = \mathsf{F}^{-1} d\mathbf{x}. \qquad (2.10)$$

These two equations show how infinitesimal *line elements* $d\mathbf{X}$ at $\mathbf{X}$ transform under the deformation into line elements $d\mathbf{x}$ at $\mathbf{x}$, and vice versa. We also note that $d\mathbf{x} = \mathsf{F} d\mathbf{X} \ne \mathbf{0}$, because a line element cannot be annihilated by a deformation. This remark shows that $\det \mathsf{F} \ne 0$ and that F must indeed be invertible.

The deformation gradient plays a major role in nonlinear elasticity. It is through F that differential operators in $\mathcal{B}_r$ are linked to their counterparts in $\mathcal{B}_c$. Knowledge of F also gives access to local changes in distances, areas, and volumes.

2.3 Differential Operators in Different Configurations

Let ϕ, $\mathbf{u}$, T be scalar, vector, and second-order tensor fields, respectively, associated with a deforming body. The following formulas prove very useful:

$$\operatorname{Grad} \phi = \mathsf{F}^\mathsf{T} \operatorname{grad} \phi, \qquad (2.11)$$

$$\operatorname{Grad} \mathbf{u} = (\operatorname{grad} \mathbf{u}) \mathsf{F}, \qquad (2.12)$$

$$\operatorname{Div} \mathbf{u} = J \operatorname{div}(J^{-1} \mathsf{F} \mathbf{u}), \qquad (2.13)$$

$$\operatorname{Div} \mathsf{T} = J \operatorname{div}(J^{-1} \mathsf{F} \mathsf{T}), \qquad (2.14)$$

where J is defined as

$$J = \det \mathsf{F}. \qquad (2.15)$$

To show the first formula (2.11), we calculate

$$\left[\mathsf{F}^\mathsf{T} \operatorname{grad} \phi\right]_i = \sum_{j=1}^{3} F^\mathsf{T}_{ij} (\operatorname{grad} \phi)_j = \sum_{j=1}^{3} F_{ji} \frac{\partial \phi}{\partial x_j} = \sum_{j=1}^{3} \sum_{k=1}^{3} F_{ji} \frac{\partial X_k}{\partial x_j} \frac{\partial \phi}{\partial X_k}$$

$$= \sum_{j=1}^{3} \sum_{k=1}^{3} F_{ji} F^{-1}_{kj} \frac{\partial \phi}{\partial X_k} = \sum_{k=1}^{3} \delta_{ki} \frac{\partial \phi}{\partial X_k} = \frac{\partial \phi}{\partial X_i} = [\operatorname{Grad} \phi]_i. \qquad (2.16)$$

The second formula is proved in the same way. The third and fourth formulas require more work and are proved later (Sect. 2.10).

2.4 Deformation of Line, Volume, and Area Elements

We saw that Equation (2.9) describes how infinitesimal *line elements* $d\mathbf{X}$ of material initially at $\mathbf{X}$ transform under the deformation into infinitesimal line elements $d\mathbf{x}$ currently at $\mathbf{x}$:

$$d\mathbf{x} = \mathbf{F} d\mathbf{X}. \tag{2.17}$$

Now we look at *volume elements* and consider the parallelepiped in $\mathcal{B}_r$ formed by line elements $d\mathbf{X}, d\mathbf{X}', d\mathbf{X}''$ at $\mathbf{X}$ (note that the prime and double prime do not denote differentiation here, simply a way of labelling different vector elements). Using (1.18), we write that its (Lagrangian) volume dV is

$$dV = d\mathbf{X} \cdot (d\mathbf{X}' \times d\mathbf{X}'') = \det\left[d\mathbf{X}\big|d\mathbf{X}'\big|d\mathbf{X}''\right], \tag{2.18}$$

where the brackets denote the matrix with the vectors $d\mathbf{X}, d\mathbf{X}', d\mathbf{X}''$ as columns (or equivalently, as rows, because $\det \mathbf{A} = \det(\mathbf{A}^\mathsf{T})$). The corresponding (Eulerian) volume dv in $\mathcal{B}_c$ is

$$\begin{aligned}dv = d\mathbf{x} \cdot (d\mathbf{x}' \times d\mathbf{x}'') &= \det\left[d\mathbf{x}\big|d\mathbf{x}'\big|d\mathbf{x}''\right] = \det\left[\mathbf{F}d\mathbf{X}\big|\mathbf{F}d\mathbf{X}'\big|\mathbf{F}d\mathbf{X}''\right] \\ &= \det\left(\mathbf{F}\left[d\mathbf{X}\big|d\mathbf{X}'\big|d\mathbf{X}''\right]\right) = \det(\mathbf{F}) \det\left[d\mathbf{X}\big|d\mathbf{X}'\big|d\mathbf{X}''\right],\end{aligned}$$

which reads

$$dv = J dV. \tag{2.19}$$

The equation relates the volume elements in one configuration to the others. It also shows that $J = \det \mathbf{F} > 0$.

From (2.19), we see that $J = \det \mathbf{F}$ is a measure of the local *changes in volume* under the deformation. When a deformation does not lead to change in volume anywhere in the solid, it is said to be *isochoric*. In that case,

$$J = \det \mathbf{F} = 1, \tag{2.20}$$

everywhere.

A solid which allows only isochoric deformations is called an *incompressible material*. In the real world, many soft solids, such as rubber, gels, and tissues, are (almost) incompressible. Looking through history, we can go back as early as 1667 for a simple proof that biological soft tissues are incompressible. Consider the ingenious experiment devised by Jan Swammerdam, a Dutch anatomist who wanted to test the existence of "animal spirits" developed by Descartes. As shown in Fig. 2.2, he put a frog thigh muscle inside an air-tight glass syringe (a) and let a nerve (c) come out through a tightly sealed hole (b). By pinching the nerve, he produced a large deformation of the muscle, and yet observed that a water bubble in

Fig. 2.2 A seventeenth-century experiment showing that soft tissues are incompressible: when the major nerve of the frog leg is pinched, the muscle contracts (large deformation) and the altitude drop of water in the thin pipette above the main chamber does not change (no volume change). (Reproduced with permission from [1])

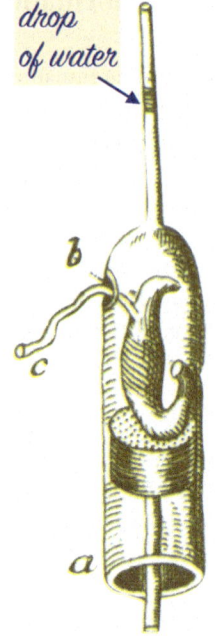

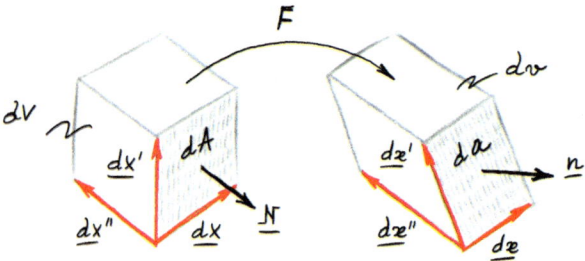

Fig. 2.3 How line, area, and volume elements deform from $\mathcal{B}_r$ to $\mathcal{B}_c$

the neck of the syringe did not change altitude. Hence, the volume had not changed, because if the muscles had lost some volume, the drop would have come down, and vice versa (incidentally, the experiment put an end to Descartes's theory of animal spirits).

Next, we look at how *surface elements* deform one into another. We call dA the area of a surface element with unit normal $\mathbf{N}$ in the reference configuration $\mathcal{B}_r$, which deforms into a surface element in the current configuration $\mathcal{B}_c$ with area da and unit normal $\mathbf{n}$; see Fig. 2.3. The surfaces elements surround a point at $\mathbf{X}$ in $\mathcal{B}_r$ and at $\mathbf{x}$ in $\mathcal{B}_c$. The line elements $d\mathbf{X}$, $d\mathbf{X}'$ and $d\mathbf{x}$, $d\mathbf{x}'$ delimit the corresponding surface elements. Then we have

$$d\mathbf{x} = \mathsf{F} d\mathbf{X}, \qquad d\mathbf{x}' = \mathsf{F} d\mathbf{X}', \qquad (2.21)$$

and the area of the parallelogram with sides $d\mathbf{X}$, $d\mathbf{X}'$ can be computed with the following cross product:

$$\mathbf{N}dA = d\mathbf{X} \times d\mathbf{X}'. \tag{2.22}$$

In $\mathcal{B}_c$, the corresponding parallelogram has sides $d\mathbf{x}$, $d\mathbf{x}'$ and area

$$\mathbf{n}da = d\mathbf{x} \times d\mathbf{x}'. \tag{2.23}$$

From (2.19), we obtain, for any third line element $d\mathbf{x}'' = \mathsf{F}d\mathbf{X}''$ (not in the $(d\mathbf{x}, d\mathbf{x}')$ plane)

$$d\mathbf{x}'' \cdot \mathbf{n}da = d\mathbf{x}'' \cdot (d\mathbf{x} \times d\mathbf{x}') = dv = JdV = Jd\mathbf{X}'' \cdot (d\mathbf{X} \times d\mathbf{X}')$$
$$= Jd\mathbf{X}'' \cdot \mathbf{N}dA = J\mathsf{F}^{-1}d\mathbf{x}'' \cdot \mathbf{N}dA = Jd\mathbf{x}'' \cdot \left(\mathsf{F}^{-1}\right)^\mathsf{T}\mathbf{N}dA, \tag{2.24}$$

where we used (1.11) for the last equality. Hence, the surface areas dA and da are related through

$$\mathbf{n}da = J(\mathsf{F}^{-1})^\mathsf{T}\mathbf{N}dA. \tag{2.25}$$

This is an important result, known as *Nanson's formula*.

2.5 Example: Simple Shear

Consider that by the application of a certain set of forces, we produce the following deformation in a soft solid:

$$x_1 = X_1 + KX_2, \qquad x_2 = X_2, \qquad x_3 = X_3, \tag{2.26}$$

where K is a constant. This deformation is called *simple shear*, and the constant K is called the *amount of shear*.

To visualise the deformation, we can find out what happens to a unit cube when its material points are displaced according to (2.26); see Fig. 2.4. We see that the deformation models what happens (ideally) when the top face of the cube is displaced parallel to the bottom face, which is held in place. The top and bottom faces remain horizontal and undeformed, with the same dimensions and areas as in the reference configuration; the front and back faces remain in the vertical plane; the side faces are slanted by the *angle of shear*, which we compute as $\tan^{-1} K$.

Fig. 2.4 A cube deformed by a simple shear of amount K

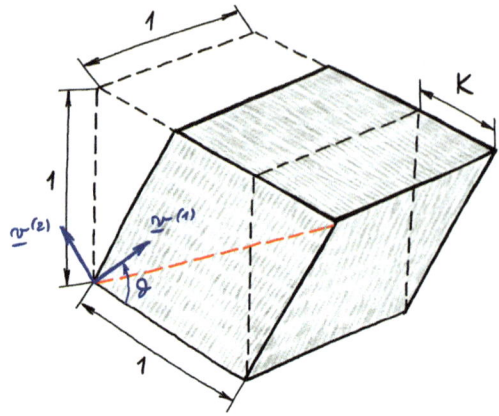

We can also compute the deformation gradient **F** to find out what happens locally to line, surface, and volume elements. Its components are $F_{ij} = \partial x_i / \partial X_j$, or

$$[\mathbf{F}] = \begin{bmatrix} 1 & K & 0 \\ 0 & 1 & 0 \\ 0 & 0 & 1 \end{bmatrix}. \tag{2.27}$$

Clearly, $J = \det \mathbf{F} = 1$, showing that there is no volume change: simple shear is an isochoric deformation.

Next, we look at how line elements change through $d\mathbf{x} = \mathbf{F}d\mathbf{X}$. Using (2.27), we see that a horizontal line element, $[d\mathbf{X}] = \begin{bmatrix} 1 \\ 0 \\ 0 \end{bmatrix} dL$, becomes $[d\mathbf{x}] = \begin{bmatrix} 1 \\ 0 \\ 0 \end{bmatrix} dL$: it remains unchanged. A vertical line element $[d\mathbf{X}] = \begin{bmatrix} 0 \\ 1 \\ 0 \end{bmatrix} dL$, becomes $[d\mathbf{x}] = \begin{bmatrix} K \\ 1 \\ 0 \end{bmatrix} dL$: it becomes tilted. Its magnitude after tilting is $d\ell = \sqrt{1 + K^2} dL$, showing that it has been stretched.

To see how surface elements change using Nanson's formula $\mathbf{n}da = J(\mathbf{F}^{-1})^{\mathsf{T}} \mathbf{N} dA$, we need to compute $\mathbf{F}^{-1}$. We find

$$[\mathbf{F}^{-1}] = \begin{bmatrix} 1 & -K & 0 \\ 0 & 1 & 0 \\ 0 & 0 & 1 \end{bmatrix}, \quad \text{so that} \quad [(\mathbf{F}^{-1})^{\mathsf{T}}] = \begin{bmatrix} 1 & 0 & 0 \\ -K & 1 & 0 \\ 0 & 0 & 1 \end{bmatrix}, \tag{2.28}$$

showing that the inverse of simple shear of amount K is simple shear of amount $-K$, as expected. Take a vertical surface element in the (X_2, X_3)-plane: $\mathbf{N}dA = \begin{bmatrix} 1 \\ 0 \\ 0 \end{bmatrix} dA$; it becomes $\mathbf{n}da = \begin{bmatrix} 1 \\ -K \\ 0 \end{bmatrix} dA$ according to Nanson's formula. Notice that the area of this surface element on the slanted face in $\mathcal{B}_c$ is $da = \sqrt{1+K^2}dA$, which is consistent with the fact that the tilted edge of this face has length $d\ell = \sqrt{1+K^2}dL$, whereas the length of the horizontal edge remains unchanged.

2.6 Composition of Deformations

Consider that a body is subjected to two successive deformations. The first one, $\mathbf{a}(\mathbf{X}) = \mathbf{x}$, brings a point at $\mathbf{X}$ in $\mathcal{B}_r$ to $\mathbf{x}$ in $\mathcal{B}_c$, with deformation gradient $\mathsf{A} = \partial \mathbf{a}/\partial \mathbf{X}$. The second one, $\mathbf{b}(\mathbf{x}) = \mathbf{x}^*$, brings a point at $\mathbf{x}$ in $\mathcal{B}_c$ to $\mathbf{x}^*$ in $\mathcal{B}_c^*$, with deformation gradient $\mathsf{B} = \partial \mathbf{b}/\partial \mathbf{x}$.

What about the combined deformation $\boldsymbol{\chi}(\mathbf{X}) = \mathbf{x}^* = \mathbf{b}(\mathbf{x}) = \mathbf{b}(\mathbf{a}(\mathbf{X}))$, bringing the point at $\mathbf{X}$ in $\mathcal{B}_r$ to $\mathbf{x}^*$ in $\mathcal{B}_c^*$? The components of its deformation gradient $\mathsf{F} = \partial \boldsymbol{\chi}/\partial \mathbf{X}$ are computed by the chain rule as

$$F_{ij} = \frac{\partial \chi_i}{\partial X_j} = \sum_{k=1}^{3} \frac{\partial b_i}{\partial x_k} \frac{\partial x_k}{\partial X_j} = \sum_{k=1}^{3} \frac{\partial b_i}{\partial x_k} \frac{\partial a_k}{\partial X_j} = \sum_{k=1}^{3} B_{ik} A_{kj}. \tag{2.29}$$

Hence, the rule of composition of deformations, when one deformation with gradient A is followed by another with gradient B, is that the combined gradient F for the whole deformation is

$$\mathsf{F} = \mathsf{BA}. \tag{2.30}$$

The order of the tensors in this decomposition is important, because in general tensors do not commute, and $\mathsf{BA} \neq \mathsf{AB}$.

In the next section, we see that any given deformation can be decomposed into a stretch followed by a rotation, or into the same rotation followed by a different stretch.

2.7 Further Results of Tensor Algebra

To analyse further the local nature of a deformation, i.e. of F, we use some properties of second-order tensors.

When we inspect the form of F for simple shear in (2.27), we see that it has an off-diagonal component. This is due to the choice of coordinates we made when we computed the components. However, as we speculated in the first chapter, there might be a way to reduce the number of non-zero components by a change of basis.

First let us look at S, a positive-definite, symmetric tensor. We call λ_i its (real, positive) eigenvalues, and **r**, **s**, **t** the corresponding unit eigenvectors. In the (**r**, **s**, **t**) orthonormal basis, the tensor is written as

$$[S] = \begin{bmatrix} \lambda_1 & 0 & 0 \\ 0 & \lambda_2 & 0 \\ 0 & 0 & \lambda_3 \end{bmatrix}, \tag{2.31}$$

which can be easily checked by writing that $\mathbf{r} = \begin{bmatrix} 1 \\ 0 \\ 0 \end{bmatrix}$ in that basis, so that $\mathbf{Sr} = \lambda_1 \mathbf{r}$ indeed, and so on. Hence, in the orthonormal basis of its eigenvectors, S is diagonal (only three non-zero components), and each element in the diagonal is an eigenvalue.

Because each eigenvalue is positive, then there exists a unique, positive-definite, symmetric tensor, U say, such that

$$\mathsf{U}^2 = \mathsf{S}. \tag{2.32}$$

It is simply the tensor U with the following components in (**r**, **s**, **t**):

$$[U] = \begin{bmatrix} \sqrt{\lambda_1} & 0 & 0 \\ 0 & \sqrt{\lambda_2} & 0 \\ 0 & 0 & \sqrt{\lambda_3} \end{bmatrix}. \tag{2.33}$$

We call U the *square root* of S.

Now the deformation gradient F is not necessarily a symmetric tensor. However, as we shall see, any deformation can be locally decomposed into a rotation, which does not alter relative distances, preceded by, or followed by, a tri-axial stretch. In other words, F is written as RU or VR, where R is a rotation and U, V are symmetric, positive-definite tensors. Remarkably, this decomposition is unique, due to the *polar decomposition theorem*, which plays a central role in continuum mechanics:

$$\mathsf{F} = \mathsf{RU} = \mathsf{VR}. \tag{2.34}$$

To prove it, we first note that the tensors FF^T and $\mathsf{F}^\mathsf{T}\mathsf{F}$ are symmetric, and positive definite (because $\mathbf{a} \cdot \mathsf{FF}^\mathsf{T}\mathbf{a} = \mathbf{a} \cdot \mathsf{F}(\mathsf{F}^\mathsf{T}\mathbf{a}) = \mathsf{F}^\mathsf{T}\mathbf{a} \cdot \mathsf{F}^\mathsf{T}\mathbf{a} = |\mathsf{F}^\mathsf{T}\mathbf{a}|^2 > 0$ for all vectors **a**, and similarly for $\mathbf{a} \cdot \mathsf{F}^\mathsf{T}\mathsf{F}\mathbf{a}$). Hence, by the square root theorem, there exist unique positive-definite symmetric tensors U, V such that

$$\mathsf{FF}^\mathsf{T} = \mathsf{V}^2, \qquad \mathsf{F}^\mathsf{T}\mathsf{F} = \mathsf{U}^2. \tag{2.35}$$

Then we introduce $R = V^{-1}F$. We see that

$$RR^T = (V^{-1}F)(V^{-1}F)^T = (V^{-1})FF^T V^{-1} = V^{-1}V^2V^{-1} = I, \qquad (2.36)$$

so that $\det R = \pm 1$. Also,

$$\det R = \det(V^{-1}F) = (\det V)^{-1}(\det F) > 0, \qquad (2.37)$$

showing that $\det R = +1$, i.e. R is a rotation. Similarly, we may define $Q = FU^{-1}$ and show that it is a rotation. Finally,

$$F = QU = (QU)(Q^T Q) = (QUQ^T)(Q), \qquad (2.38)$$

where clearly, the first bracketed tensor is symmetric and the second is a rotation. By uniqueness of the decomposition $F = VR$, it follows that $V = RUR^T$ and $Q = R$.

Now let us call λ_1, λ_2, λ_3 the eigenvalues of U, with corresponding unit eigenvectors $\mathbf{u}^{(1)}$, $\mathbf{u}^{(2)}$, $\mathbf{u}^{(3)}$. The eigenvalues are real positive and the eigenvectors are orthogonal, because U is symmetric positive definite; see Sect. 1.7.

Then introduce the vectors $\mathbf{v}^{(i)} = R\mathbf{u}^{(i)}$. They are unit vectors because, for example, $\mathbf{v}^{(1)} \cdot \mathbf{v}^{(1)} = R\mathbf{u}^{(1)} \cdot R\mathbf{u}^{(1)} = \mathbf{u}^{(1)} \cdot R^T R \mathbf{u}^{(1)} = \mathbf{u}^{(1)} \cdot \mathbf{u}^{(1)} = 1$. Also, we note that

$$V\mathbf{v}^{(1)} = VR\mathbf{u}^{(1)} = F\mathbf{u}^{(1)} = RU\mathbf{u}^{(1)} = \lambda_1 R\mathbf{u}^{(1)} = \lambda_1 \mathbf{v}^{(1)}, \qquad (2.39)$$

and similarly for the other eigenvectors.

We conclude that U and V share the same eigenvalues λ_i but have different eigenvectors, $\mathbf{u}^{(i)}$ and $\mathbf{v}^{(i)} = R\mathbf{u}^{(i)}$, respectively.

We introduce the tensors

$$C = F^T F = U^2, \qquad B = FF^T = V^2. \qquad (2.40)$$

called the *right* and *left Cauchy-Green deformation tensors*, respectively. They are both symmetric, positive-definite tensors. We see that the unit eigenvectors $\mathbf{u}^{(i)}$ of U are also the unit eigenvectors of C, but with eigenvalues λ_i^2, because $C\mathbf{u}^{(i)} = UU\mathbf{u}^{(i)} = \lambda_i U\mathbf{u}^{(i)} = \lambda_i^2 \mathbf{u}^{(i)}$. Similarly, the unit eigenvectors $\mathbf{v}^{(i)}$ of V are also the unit eigenvectors of B, with corresponding eigenvalues λ_i^2.

2.8 Stretch and Strain of Line Elements

We just saw that the deformation gradient F can be uniquely decomposed into a rotation R, preceded (or followed) by a stretch U (or V). We now explain why U and V are called *stretch tensors*.

We call **M** a unit vector along the infinitesimal line element $d\mathbf{X}$ lying in the reference configuration and **m** the unit vector along the deformed line element $d\mathbf{x}$ in $\mathcal{B}_c$. So, $d\mathbf{X} = \mathbf{M}|d\mathbf{X}|$, $d\mathbf{x} = \mathbf{m}|d\mathbf{x}|$ and (2.9) gives $\mathbf{m}|d\mathbf{x}| = d\mathbf{x} = \mathsf{F}d\mathbf{X} = \mathsf{F}\mathbf{M}|d\mathbf{X}|$. It follows that

$$|d\mathbf{x}|^2 = d\mathbf{x} \cdot d\mathbf{x} = (\mathsf{F}\mathbf{M}) \cdot (\mathsf{F}\mathbf{M})|d\mathbf{X}|^2 = (\mathsf{F}^\mathsf{T}\mathsf{F}\mathbf{M}) \cdot \mathbf{M}|d\mathbf{X}|^2, \qquad (2.41)$$

and we find that the local change in length from $\mathcal{B}_r$ to $\mathcal{B}_c$ can be computed as the ratio

$$\lambda(\mathbf{M}) = \frac{|d\mathbf{x}|}{|d\mathbf{X}|} = |\mathsf{F}\mathbf{M}| = \sqrt{\mathbf{M} \cdot \mathsf{F}^\mathsf{T}\mathsf{F}\mathbf{M}} = \sqrt{\mathbf{M} \cdot \mathsf{C}\mathbf{M}}. \qquad (2.42)$$

We call $\lambda(\mathbf{M})$ the *stretch in the direction* **M** at **X**. The *extension* is the quantity $e = \lambda - 1 = (|d\mathbf{x}| - |d\mathbf{X}|)/|d\mathbf{X}|$, giving the relative change in length.

Because U is positive definite and symmetric, there exist unit, orthogonal eigenvectors $\mathbf{u}^{(i)}$ such that its components in the $(\mathbf{u}^{(1)}, \mathbf{u}^{(2)}, \mathbf{u}^{(3)})$ basis are

$$[\mathsf{U}] = \begin{bmatrix} \lambda_1 & 0 & 0 \\ 0 & \lambda_2 & 0 \\ 0 & 0 & \lambda_3 \end{bmatrix}, \qquad (2.43)$$

where $\lambda_i > 0$ are called the *principal stretches* of the deformation, and $\mathbf{u}^{(i)}$ are along the *principal axes*.

An important property of the principal stretches is that when they are ordered as $\lambda_1 \geq \lambda_2 \geq \lambda_3$, then λ_1 and λ_3 define the *largest* and *smallest* stretches of all line elements in the body, respectively. In other words, if **M** is a unit vector in the reference configuration, then

$$\lambda_1 \geq \lambda(\mathbf{M}) \geq \lambda_3 \qquad \text{for all } \mathbf{M}. \qquad (2.44)$$

To prove it, decompose the unit vector **M** in the basis $(\mathbf{u}^{(1)}, \mathbf{u}^{(2)}, \mathbf{u}^{(3)})$ of the principal directions of U, so that $\mathbf{M} = M_1 \mathbf{u}^{(1)} + M_2 \mathbf{u}^{(2)} + M_3 \mathbf{u}^{(3)}$. Then, it is clear that

$$\lambda^2(\mathbf{M}) = \mathbf{M} \cdot \mathsf{C}\mathbf{M} = M_1^2 \lambda_1^2 + M_2^2 \lambda_2^2 + M_3^2 \lambda_3^2. \qquad (2.45)$$

Now, because $\lambda_1 \geq \lambda_2 \geq \lambda_3$ and $M_1^2 + M_2^2 + M_3^2 = 1$, we find

$$\lambda_1^2 \geq \lambda^2(\mathbf{M}) \geq \lambda_3^2 \qquad \text{for all } \mathbf{M}, \qquad (2.46)$$

which gives the desired result.

The tensors U and V are called the *right* and *left stretch tensors*, respectively. The deformation, as tracked by the deformation tensor F, rotates the principal axes of U into those of V and effects stretching along those directions. The principal axes of U

(and thus, of $C = U^2$) are the *Lagrangian principal axes* and the principal axes of V (and thus, of $B = V^2$) are the *Eulerian principal axes*. Finding the eigenvectors and eigenvalues of V (and thus, of B) gives the directions in which a solid is most and least stretched in $\mathcal{B}_c$, and by which amounts.

Strain is measured locally by changes in the lengths of line elements. In other words, strain measures the changes in distance of two neighbouring particles, one at position $\mathbf{X}$, the other at position $\mathbf{X} + d\mathbf{X}$, in the reference configuration, mapped into positions $\mathbf{x}$ and $\mathbf{x} + d\mathbf{x}$ in the current configuration.

For example, the quantity $|d\mathbf{x}|^2 - |d\mathbf{X}|^2 = d\mathbf{X} \cdot (\mathsf{F}^\mathsf{T}\mathsf{F} - \mathsf{I})d\mathbf{X}$ gives the change in the squared length of a line element. Thus, the tensor $\mathsf{F}^\mathsf{T}\mathsf{F} - \mathsf{I}$ is a *measure of strain*. The so-called *Green strain tensor* E is defined by

$$\mathsf{E} = \tfrac{1}{2}(\mathsf{F}^\mathsf{T}\mathsf{F} - \mathsf{I}). \tag{2.47}$$

Then we see that the squared lengths of line elements have changed as

$$|d\mathbf{x}|^2 - |d\mathbf{X}|^2 = 2(d\mathbf{X}) \cdot \mathsf{E}(d\mathbf{X}) = 2E_{ij}dX_i dX_j, \tag{2.48}$$

and that E refers to (squared) length changes with respect to the material (Lagrangian) line elements.

Conversely, we can express the changes with respect to the spatial (Eulerian) line elements, as

$$|d\mathbf{x}|^2 - |d\mathbf{X}|^2 = 2d\mathbf{x} \cdot \mathsf{e}d\mathbf{x}, \tag{2.49}$$

where

$$\mathsf{e} = \tfrac{1}{2}\left[\mathsf{I} - (\mathsf{F}^{-1})^\mathsf{T}\mathsf{F}^{-1}\right] \tag{2.50}$$

is called the *Eulerian strain tensor*.

Note that the Green strain tensor E may be written as

$$\mathsf{E} = \tfrac{1}{2}(\mathsf{C} - \mathsf{I}) = \tfrac{1}{2}(\mathsf{U}^2 - \mathsf{I}), \tag{2.51}$$

and the Eulerian strain tensor e may be written as

$$\mathsf{e} = \tfrac{1}{2}(\mathsf{I} - \mathsf{B}^{-1}) = \tfrac{1}{2}(\mathsf{I} - \mathsf{V}^{-2}). \tag{2.52}$$

2.9 Homogeneous Deformations

When F is independent of $\mathbf{X}$ (i.e. when all the F_{ij} are constants), the deformation is said to be *homogeneous*. Homogeneous deformations are quite easy to implement

Fig. 2.5 Two examples of homogeneous deformations of a cube (dashed in the reference configuration): simple elongation along the three directions with $\lambda_1 > 1$, and pure dilatation of a compressible solid, with $\lambda < 1$

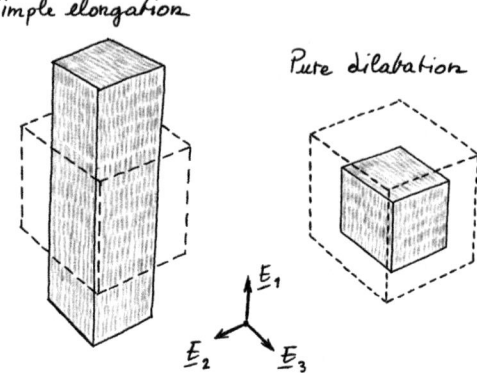

experimentally, and they thus form the basis for the standardised testing and evaluation of the elastic properties of solids. The following examples are important and practical cases of homogeneous deformations.

2.9.1 Simple Elongation

Consider the uniform axial extension of a solid thin strip along the $\mathbf{E}_1$ direction, say (Fig. 2.5). The strip is deformed into a longer strip, and there is no rotation of the principal axes of stretch: $\mathsf{R} = \mathsf{I}$, $\mathsf{F} = \mathsf{U} = \mathsf{V}$, $\mathbf{u}^{(i)} = \mathbf{v}^{(i)} = \mathbf{E}_i = \mathbf{e}_i$. The axial stretch λ_1 is linked to the elongation, also called extension, through $e = \lambda_1 - 1$. Because there is symmetry in the plane perpendicular to the elongation, the principal stretches λ_2 and λ_3 must be equal: $\lambda_2 = \lambda_3$ (strictly speaking, that is only true when the material is "isotropic" in the cross-section plane, a notion that we elucidate in Chap. 4).

The deformation gradient has the following components:

$$[\mathsf{F}] = [\mathsf{U}] = \begin{bmatrix} \lambda_1 & 0 & 0 \\ 0 & \lambda_2 & 0 \\ 0 & 0 & \lambda_2 \end{bmatrix} = \left[\frac{\partial x_i}{\partial X_j} \right]. \quad (2.53)$$

The corresponding deformation, *simple elongation*, is thus

$$x_1 = \lambda_1 X_1, \qquad x_2 = \lambda_2 X_2, \qquad x_3 = \lambda_2 X_3. \quad (2.54)$$

Hence, a strip with material length L and cross-sectional area A is deformed into strip of spatial length ℓ and cross-section a, with the connections

$$\ell = \lambda_1 L, \qquad a = \lambda_2^2 A. \quad (2.55)$$

For *incompressible* solids, we have $\det F = 1$ everywhere, so that $\lambda_1 \lambda_2^2 = 1$, giving $\lambda_2 = \lambda_1^{-1/2}$. Hence, an elongated ($\lambda_1 > 1$) incompressible rod contracts laterally ($\lambda_2 = \lambda_1^{-1/2} < 1$).

2.9.2 Pure Dilatation

A soft solid is under pure dilatation when it is stretched by the same amount in all directions (Fig. 2.5). All principal stretches are equal, and we call $\lambda = \lambda_1 = \lambda_2 = \lambda_3$, so that $F = \lambda I$. In dilatation, a cube is deformed into another cube and a sphere into another sphere, etc.; in other words, dilatation is volume-changing and shape-preserving.

Here, incompressibility imposes $\lambda^3 = 1$, that is, $\lambda = 1$: incompressible materials cannot undergo pure dilatation.

2.9.3 Simple Shear

As we saw earlier, simple shear is defined by the deformation

$$x_1 = X_1 + KX_2, \qquad x_2 = X_2, \qquad x_3 = X_3, \qquad (2.56)$$

where K, a constant, is called the *amount of shear* and $\tan^{-1} K$ is the *angle of shear*.

Take a look at Fig. 2.6, which shows a cuboid of porcine brain matter being subjected to a simple shear of amount $K = 1$ (angle of shear is thus $45°$). Along

Fig. 2.6 Simple shear testing of porcine brain matter. The sample is a cuboid with dimensions $19 \times 19 \times 7$ mm and has two opposite faces glued to parallel platens, the top one fixed and the bottom one mobile. The middle picture is taken after a bottom platen displacement of 7 mm, corresponding to the amount of shear $K = 1.0$, and the angle of shear $45°$. (Reproduced with permission from [2])

which direction is it being stretched the most, and by which amount? Our intuition tells us that the largest stretch should occur along the long diagonal, but to answer this question precisely, we need to determine the direction of the Eulerian principal axes.

Using the coordinate systems in (2.56), we find that the components of the deformation gradient are

$$[\mathbf{F}] = \begin{bmatrix} 1 & K & 0 \\ 0 & 1 & 0 \\ 0 & 0 & 1 \end{bmatrix}. \tag{2.57}$$

To find the Eulerian principal axes, we first compute

$$[\mathbf{V}^2] = [\mathbf{F}\mathbf{F}^T] = \begin{bmatrix} 1+K^2 & K & 0 \\ K & 1 & 0 \\ 0 & 0 & 1 \end{bmatrix}. \tag{2.58}$$

To find the principal stretches, we compute the eigenvalues λ^2 of $\mathbf{V}^2$ by solving

$$\det(\mathbf{V}^2 - \lambda^2 \mathbf{I}) = 0, \tag{2.59}$$

or

$$\begin{vmatrix} 1+K^2-\lambda^2 & K & 0 \\ K & 1-\lambda^2 & 0 \\ 0 & 0 & 1-\lambda^2 \end{vmatrix} = 0, \tag{2.60}$$

which factorises as

$$(\lambda^2 - 1)[\lambda^4 - (2+K^2)\lambda^2 + 1] = 0. \tag{2.61}$$

Let the positive roots of the quartic (in the squared brackets) be denoted by λ_1, λ_2. The third positive root is clearly $\lambda_3 = 1$ (round parentheses). Then, looking at the quartic in λ as a quadratic in λ^2, we see that the sum and product of the squared roots are

$$\lambda_1^2 + \lambda_2^2 = 2 + K^2, \qquad \lambda_1^2 \lambda_2^2 = 1. \tag{2.62}$$

Because $\lambda_1 \lambda_2 = 1$, one of these eigenvalues is greater than 1 and the other is smaller than 1. Now call λ_1 the eigenvalue greater than 1; then $\lambda_2 = \lambda_1^{-1} < 1$, and the equation above reads

$$\lambda_1^2 + \lambda_1^{-2} = 2 + K^2, \quad \text{or} \quad K^2 = \lambda_1^2 - 2 + \lambda_1^{-2} = (\lambda_1 - \lambda_1^{-1})^2, \tag{2.63}$$

2.9 Homogeneous Deformations

and hence

$$K = \lambda_1 - \lambda_1^{-1} = \lambda_2^{-1} - \lambda_2, \qquad (2.64)$$

taking $K \geq 0$ to correspond to $\lambda_1 \geq 1$. We see that these equations are quadratics in λ_1 and λ_2, which we may solve explicitly. It follows that the eigenvalues of $\mathbf{V}$ are

$$\lambda_1 = \frac{K}{2} + \sqrt{1 + \frac{K^2}{4}}, \qquad \lambda_2 = -\frac{K}{2} + \sqrt{1 + \frac{K^2}{4}}, \qquad \lambda_3 = 1. \qquad (2.65)$$

Thus, the brain sample in Fig. 2.6 is experiencing a maximum stretch of $\lambda_1 = 0.5 + \sqrt{1.25} \simeq 1.62$, that is, a 62% extension, and least stretch of $\lambda_2 = -0.5 + \sqrt{1.25} \simeq 0.62$, that is, a 38% contraction. Simple shear is quite a practical protocol to obtain large stretches with very fragile soft solids such as brain matter, which would not be easy to clamp in tensile testing, for example.

Now let us determine along which directions these principal stretches occur, by looking for the principal axes, which are along the eigenvectors $\mathbf{v}^{(i)}$ of $\mathbf{V}$.

We call θ the angle between x_1 and $\mathbf{v}^{(1)}$; see Fig. 2.4, and we solve $\mathbf{V}^2 \mathbf{v}^{(1)} = \lambda_1^2 \mathbf{v}^{(1)}$, or

$$\begin{bmatrix} 1 + K^2 & K & 0 \\ K & 1 & 0 \\ 0 & 0 & 1 \end{bmatrix} \begin{bmatrix} \cos\theta \\ \sin\theta \\ 0 \end{bmatrix} = \lambda_1^2 \begin{bmatrix} \cos\theta \\ \sin\theta \\ 0 \end{bmatrix}. \qquad (2.66)$$

Taking the second line, for example, leads to

$$\tan\theta = \frac{K}{\lambda_1^2 - 1} = \frac{K}{\left(\frac{K}{2}\right)^2 + K\sqrt{1 + \frac{K^2}{4}} + 1 + \frac{K^2}{4} - 1} = \frac{1}{\frac{K}{2} + \sqrt{1 + \frac{K^2}{4}}} = \frac{1}{\lambda_1}. \qquad (2.67)$$

Hence, we find that

$$\tan\theta = \lambda_2 = -\frac{K}{2} + \sqrt{1 + \frac{K^2}{4}}. \qquad (2.68)$$

Finally, consider the cubic sample of Fig. 2.4 of side length and height $L = H$, subject to simple shear of amount K. The diagonal is at the angle α such that $\tan\alpha = 1/(1 + K)$. Clearly, $\alpha \neq \theta$ and the greatest stretch does not occur along the diagonal, contrary to what we might have expected intuitively. Of course, for blocks with other aspect ratios L/H, the diagonal is at a different angle, but the conclusion remains the same in general. It is then a simple exercise to show that $\alpha = \theta$ only when the block is sheared by the specific amount $K = H/L - L/H$.

2.9.4 Shear Box Deformation

Consider a cuboid sample of a soft solid, placed inside a cubic box made of four rigid faces, for which the four edges parallel to $\mathbf{E}_3$ are hinged. Pushing the box in the (X_1, X_2) plane produces the *shear box deformation*. Provided the faces of the sample are well lubricated, the deformation is reasonably described by

$$x_1 = X_1 + (\sin\phi)X_2, \qquad x_2 = (\cos\phi)X_2, \qquad x_3 = \lambda_3 X_3,$$

where λ_3 is a constant; see Fig. 2.7 for the definition of the shear box angle ϕ.

To find this expression for the deformation, we tracked the locations of the vertices of the vertical squares in the reference configuration into those of the rhombus in the current deformation. The way the cube deforms out-of-plane is not easy to find geometrically, so we simply wrote that there is a yet unknown stretch λ_3 in that direction.

Then we conduct tests on jelly, a very soft material with a large water content, which makes it incompressible. This assumption helps us find λ_3, because the deformation gradient is

$$[\mathsf{F}] = \begin{bmatrix} 1 & \sin\phi & 0 \\ 0 & \cos\phi & 0 \\ 0 & 0 & \lambda_3 \end{bmatrix}, \qquad (2.69)$$

and imposing $\det \mathsf{F} = 1$ gives $\lambda_3 = 1/\cos\phi$.

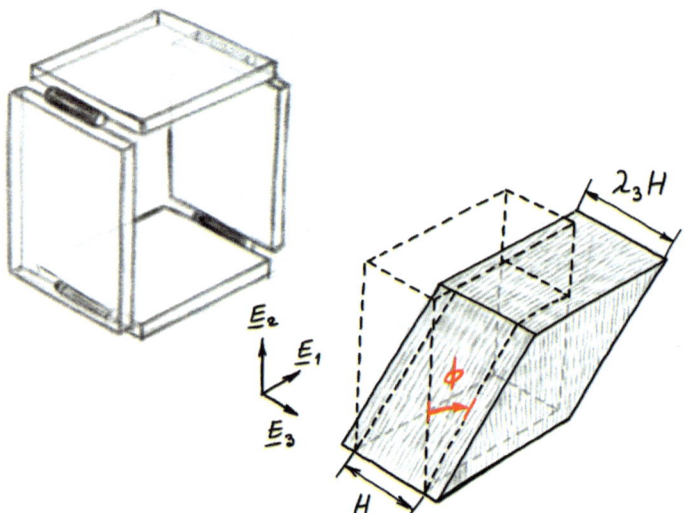

Fig. 2.7 The shear box deformation

2.9 Homogeneous Deformations

Fig. 2.8 Large shear box deformation of a cubic block of strawberry jelly. Eventually creases (left, on fresh jelly) or wrinkles (right, on older jelly where a stiffer film has formed on the surface) appear. They seem to be aligned with the large diagonal of the shear box, at right-angle to the short diagonal. (Reproduced with permission from [3])

The photos in Fig. 2.8 show what happens when ϕ is large. For fresh jelly (left), fracture lines (creases) develop aligned with the long diagonal (see yellow arrows). For older jelly (right), a film has formed on the top surface and smooth wrinkles appear, aligned with the long diagonal. These observations suggest that the direction of the greatest contraction is along the short diagonal of the sheared box. Let us investigate if that is indeed the case.

To find the Eulerian principal axes, we compute

$$[\mathbf{V}^2] = [\mathbf{F}\mathbf{F}^\mathsf{T}] = \begin{bmatrix} 1 + \sin^2\phi & \cos\phi\sin\phi & 0 \\ \cos\phi\sin\phi & \cos^2\phi & 0 \\ 0 & 0 & \frac{1}{\cos^2\phi} \end{bmatrix}. \tag{2.70}$$

Finding the eigenvectors of $\mathbf{V}^2$ is laborious, so here we take a shortcut and use a computer algebra system. It tells us that the eigenvalues are $\lambda_1^2 = 1 + \sin\phi$, $\lambda_2^2 = 1 - \sin\phi$, $\lambda_3^2 = 1/\cos^2\phi$, and that the eigenvectors associated with the contraction stretch λ_2 are parallel to the vector

$$\mathbf{v} = \begin{bmatrix} -\cos\phi \\ 1 + \sin\phi \\ 0 \end{bmatrix}. \tag{2.71}$$

Now let us find a vector **d** aligned with the long diagonal. It makes an angle $\frac{1}{2}(\frac{\pi}{2} - \phi)$ with the horizontal, see Fig. 2.7, so that

$$\mathbf{d} = \begin{bmatrix} \cos\frac{1}{2}(\frac{\pi}{2} - \phi) \\ \sin\frac{1}{2}(\frac{\pi}{2} - \phi) \\ 0 \end{bmatrix} = \frac{1}{\sqrt{2}} \begin{bmatrix} \sqrt{1 + \cos(\frac{\pi}{2} - \phi)} \\ \sqrt{1 - \cos(\frac{\pi}{2} - \phi)} \\ 0 \end{bmatrix} = \frac{1}{\sqrt{2}} \begin{bmatrix} \sqrt{1 + \sin\phi} \\ \sqrt{1 - \sin\phi} \\ 0 \end{bmatrix}, \tag{2.72}$$

where we used the half-angle and complementary angle trigonometric identities. Now we see that

$$\mathbf{d} = \frac{1}{\sqrt{2}\sqrt{1 + \sin\phi}} \begin{bmatrix} 1 + \sin\phi \\ \sqrt{1 - \sin^2\phi} \\ 0 \end{bmatrix} = \frac{1}{\sqrt{2}\sqrt{1 + \sin\phi}} \begin{bmatrix} 1 + \sin\phi \\ \cos\phi \\ 0 \end{bmatrix}, \tag{2.73}$$

is clearly perpendicular to **v**.

Because the short diagonal is perpendicular to the long one in a rhombus, it means that the wavefronts of the creases/wrinkles, occurring at right angle to the direction of greatest contraction, are indeed aligned with the short diagonal of the shear box.

2.10 Divergence Theorem

Here we collect some results from vector and tensor calculus that are needed in the next chapters.

The *divergence theorem* for a vector field **u** reads

$$\int_R \operatorname{div} \mathbf{u} \, dv = \int_{\partial R} \mathbf{u} \cdot \mathbf{n} \, da, \tag{2.74}$$

where R is a domain in space and ∂R is its boundary, which is a closed surface.

Now take $\mathbf{u} = \phi \mathbf{b}$, where ϕ is a scalar field and **b** is an arbitrary constant vector. We have $\operatorname{div}(\phi \mathbf{b}) = (\operatorname{grad}\phi) \cdot \mathbf{b}$ because $\operatorname{div} \mathbf{b} = 0$, and we conclude that

$$\int_R \operatorname{grad}\phi \, dv = \int_{\partial R} \phi \mathbf{n} \, da, \tag{2.75}$$

or, in index notation,

$$\int_R \frac{\partial \phi}{\partial x_k} dv = \int_{\partial R} \phi n_k \, da. \tag{2.76}$$

2.10 Divergence Theorem

This equation applies to the components T_{ij} of any tensor:

$$\int_R \frac{\partial T_{ij}}{\partial x_k} dv = \int_{\partial R} T_{ij} n_k da. \qquad (2.77)$$

Writing this equation when $i = k = 1, 2, 3$ and summing gives

$$\int_R \sum_{i=1}^{3} \frac{\partial T_{ij}}{\partial x_i} dv = \int_{\partial R} \sum_{i=1}^{3} T_{ij} n_i da \qquad (2.78)$$

or, in tensor notation,

$$\int_R \text{div}\, \mathbf{T}\, dv = \int_{\partial R} \mathbf{T}^\mathsf{T} \mathbf{n}\, da. \qquad (2.79)$$

This formula provides nice proofs for (2.13) and (2.14). Hence, we have

$$\int_{R_r} \text{Div}\, \mathbf{T}\, dV = \int_{\partial R_r} \mathbf{T}^\mathsf{T} \mathbf{N}\, dA = \int_{\partial R_c} \mathbf{T}^\mathsf{T} J^{-1} \mathbf{F}^\mathsf{T} \mathbf{n}\, da$$

$$= \int_{\partial R_c} \left(J^{-1} \mathbf{F} \mathbf{T}\right)^\mathsf{T} \mathbf{n}\, da = \int_{R_c} \text{div}\left(J^{-1} \mathbf{F} \mathbf{T}\right) dv = \int_{R_r} J \text{div}\left(J^{-1} \mathbf{F} \mathbf{T}\right) dV, \qquad (2.80)$$

where we used (2.79) twice and Nanson's formula once. Because this identity is valid for arbitrary domains R_r, (2.14) follows. The proof of (2.13) follows the same lines, starting from (2.74).

Another application of the divergence theorem is in the proof of the identity

$$\text{Div}(J\mathbf{F}^{-1}) = \mathbf{0}. \qquad (2.81)$$

Take any closed region R_c in the current configuration. Then $\int_{\partial R_c} \mathbf{n}\, da = \int_{R_c} (\text{div}\, \mathbf{I}) dv = \mathbf{0}$, and using Nanson's formula, we have

$$\mathbf{0} = \int_{\partial R_c} \mathbf{n}\, da = \int_{\partial R_r} J(\mathbf{F}^{-1})^\mathsf{T} \mathbf{N}\, dA = \int_{R_r} \text{Div}\left(J\mathbf{F}^{-1}\right) dV, \qquad (2.82)$$

which holds for all closed regions, thus leading to the sought identity. Similarly, we may start from $\int_{\partial R_r} \mathbf{N}\, dA = \int_{R_r} (\text{Div}\, \mathbf{I}) dV = \mathbf{0}$ for any closed region in the reference configuration, to arrive at at the identity

$$\text{div}(J^{-1}\mathbf{F}) = \mathbf{0}. \qquad (2.83)$$

References

1. M. Cobb. Exorcizing the animal spirits: Jan Swammerdam on nerve function. *Nature Reviews Neuroscience*, 3 (2002) 395–400.
2. M. Destrade, M.D. Gilchrist, J.G. Murphy, B. Rashid, G. Saccomandi. Extreme softness of brain matter in simple shear. *International Journal of Non-Linear Mechanics*, 75 (2015) 54–58.
3. M. Carfagna, M. Destrade, A.L. Gower, A. Grillo. Oblique wrinkles. *Philosophical Transactions of the Royal Society A*, 375 (2017) 20160158.

Chapter 3
Stress

With this chapter, we see that writing the equations of equilibrium for soft solids requires more work than for rigid bodies, where the balance of linear and angular momentum of external forces and torques is sufficient to secure equilibrium. For soft solids, these conditions are only necessary; they are not sufficient for equilibrium. This calls for the introduction of internal actions, boiling down to the definition of the *stress tensor*, a fundamental player of elasticity, and the main character of this chapter.

3.1 Balance Equations

The forces acting on a continuous body can be divided into two classes: *body forces* and *contact forces*. The former type describes interactions at a distance (such as forces arising from gravity or electromagnetic fields, for example). The latter type describes interactions that require contact, either between mutual parts of a body (*internal* contact forces) or between the body and its surrounding environment (*external* contact forces).

While external contact forces and distance forces are already familiar from the study of rigid body mechanics, internal contact forces arise specifically in the study of continuous deformable bodies. Of special interest here are *surface* contact forces, which describe the mechanical interaction between adjacent parts of a body across an ideal cut surface.

For the sake of simplicity, in this chapter, we neglect body forces, even though in the final chapter we show that some special types of body forces (such as those arising from electric fields) may also be described in terms of contact forces.

A mathematical description of surface contact forces relies on *Cauchy stress principle*, which is regarded as an axiom. It can be formulated as follows, with reference to Fig. 3.1.

Fig. 3.1 The surface traction vector **t**(**x**, **n**), at a point **x** and relative to the orientation **n**, represents the force per unit area exerted by $\mathcal{B}_c^+$ on $\mathcal{B}_c^-$ across the plane $\mathcal{S}_c$

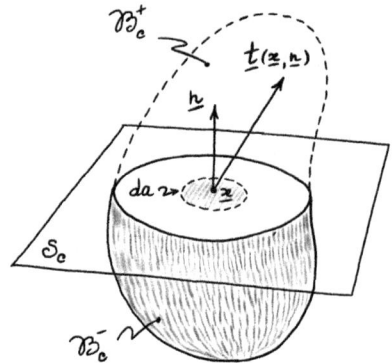

Consider a material body $\mathcal{B}_c$ and a plane $\mathcal{S}_c$ separating it into two regions $\mathcal{B}_c^+$ and $\mathcal{B}_c^-$. Also consider a point **x** on $\mathcal{S}_c$ and denote by **n** the normal to $\mathcal{S}_c$ oriented from $\mathcal{B}_c^-$ to $\mathcal{B}_c^+$. Then the force per unit area that $\mathcal{B}_c^+$ exerts on $\mathcal{B}_c^-$ at **x** is a vector field **t**(**x**, **n**), called the *surface traction*.

The surface traction is a fundamental character of Continuum Mechanics. Its dependence on the orientation **n** is the subject of the main theorem discussed in this chapter, namely *Cauchy's Theorem*. For simplicity of notation, from now on the surface traction is denoted **t**(**n**), keeping in mind that there is also a dependence on **x**. The infinitesimal force exerted across the area da at **x** is then $d\mathbf{f} = \mathbf{t}(\mathbf{n})\, da$.

Make now reference to Fig. 3.2a, where a subregion $\mathcal{R}_c$ of the body $\mathcal{B}_c$ is highlighted. If **t**(**n**) denotes the surface traction exerted on $\mathcal{R}_c$ by the remainder of the body, the resultant force exerted upon $\mathcal{R}_c$ may be computed as

$$\mathbf{f}(\mathcal{R}_c) = \int_{\partial \mathcal{R}_c} \mathbf{t}(\mathbf{n})\, da. \tag{3.1}$$

Likewise, the resultant moment exerted upon $\mathcal{R}_c$ relative to the origin is

$$\mathbf{m}_o(\mathcal{R}_c) = \int_{\partial \mathcal{R}_c} \mathbf{x} \times \mathbf{t}(\mathbf{n})\, da. \tag{3.2}$$

We are now in a position to enunciate *Euler's Axioms*, according to which a deformable body is in equilibrium if and only if the resultant force and resultant moment exerted upon any of its parts are zero,

$$\mathbf{f}(\mathcal{R}_c) = \mathbf{0}, \qquad \mathbf{m}_o(\mathcal{R}_c) = \mathbf{0}, \tag{3.3}$$

for *all* sub-parts $\mathcal{R}_c$ of $\mathcal{B}_c$. Note that these equations are more general than the requirements of equilibrium for a rigid body, where (3.3) must hold only for $\mathcal{R}_c = \mathcal{B}_c$.

3.2 The Theory of Stress: Cauchy's Theorem

We now see that to characterise fully the surface traction **t(n)** at any point and relative to any orientation **n**, it is enough to know six scalars, collected into the so-called *stress tensor*: the proof of its existence constitutes the first point of *Cauchy's Theorem*. The second and third points of the theorem establish local forms of equilibrium for a continuous body.

To introduce the stress tensor, we take a look at a region $\mathcal{R}_c$ inside the solid, and call A its surface area; see Fig. 3.2b. Consider a cross-section of $\mathcal{R}_c$, with area D. Then the region $\mathcal{R}_c$ can be seen as the two regions glued together, with areas $B + D$ for the upper part, and $C + D$ for the lower part. Consider an elementary area da in D: it has outward normal **n** when it is considered as belonging to the lower part, and outward normal $-$**n** when it is considered as belonging to the upper part.

It follows that

$$\begin{aligned}
\mathbf{0} = \int_A \mathbf{t(m)}da &= \int_C \mathbf{t(m)}da + \int_D \mathbf{t(n)}da + \int_D \mathbf{t(-n)}da + \int_B \mathbf{t(m)}da \\
&= \int_{B+C} \mathbf{t(m)}da + \int_D [\mathbf{t(n)} + \mathbf{t(-n)}]da \\
&= \int_A \mathbf{t(m)}da + \int_D [\mathbf{t(n)} + \mathbf{t(-n)}]da \\
&= \mathbf{0} + \int_D [\mathbf{t(n)} + \mathbf{t(-n)}]da.
\end{aligned} \qquad (3.4)$$

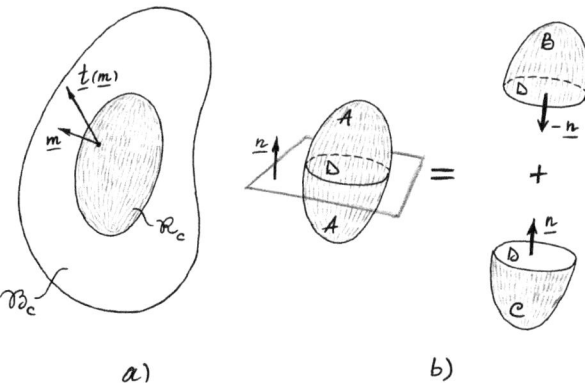

a) b)

Fig. 3.2 (**a**) $\mathcal{R}_c$ is an internal part of the current configuration of the body $\mathcal{B}_c$. The surface traction vector **t(m)** denotes the action of $\mathcal{B}_c \backslash \mathcal{R}_c$ on $\mathcal{R}_c$. (**b**) The part $\mathcal{R}_c$ may be cut through a plane $\mathcal{S}_c$ into sub-parts $\mathcal{R}_c^+$ and $\mathcal{R}_c^-$

Fig. 3.3 A small tetrahedron and its decomposition into four faces $\mathcal{A}_n, \mathcal{A}_1, \mathcal{A}_2, \mathcal{A}_3$ with areas A_n, A_1, A_2, A_3 and outward unit normals $\mathbf{n}, -\mathbf{e}_1, -\mathbf{e}_2, -\mathbf{e}_3$, respectively

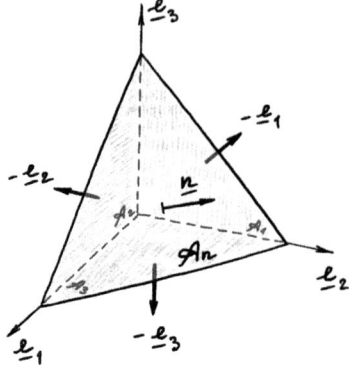

Because $\mathcal{R}_c$ is an arbitrary region, and therefore also the surface D is arbitrary, we conclude that

$$\mathbf{t}(\mathbf{n}) = -\mathbf{t}(-\mathbf{n}). \tag{3.5}$$

This is *Cauchy's Lemma*. It essentially corresponds to Newton's Third Law of "Action and Reaction" for surface tractions. It states that surface tractions acting on opposite sides of the same surface are equal in magnitude and opposite in direction.

Now we look at a tetrahedron made of four faces: a slanted face $\mathcal{A}_n$, with unit normal $\mathbf{n} = (n_1, n_2, n_3)$ and area A, and the three faces $\mathcal{A}_1, \mathcal{A}_2, \mathcal{A}_3$ with unit normals $-\mathbf{e}_1, -\mathbf{e}_2, -\mathbf{e}_3$ and areas A_1, A_2, A_3, respectively; see Fig. 3.3.

Before going further, recall that by the divergence theorem, for any region $\mathcal{R}_c$ in the current configuration with boundary surface $\partial \mathcal{R}_c$ and normal $\mathbf{m}$, the following identity holds:

$$\int_{\partial \mathcal{R}_c} \mathbf{m}\, da = \int_{\mathcal{R}_c} (\text{div } \mathbf{I})\, dv = \mathbf{0}. \tag{3.6}$$

Take now the region $\mathcal{R}_c$ to coincide with the tetrahedron described in Fig. 3.3. Then, taking the dot product of the identity (3.6) by $\mathbf{e}_2$, for example, we have

$$\mathbf{e}_2 \cdot \left(\int_{\mathcal{A}_n} \mathbf{n}\, da - \int_{\mathcal{A}_1} \mathbf{e}_1\, da - \int_{\mathcal{A}_2} \mathbf{e}_2\, da - \int_{\mathcal{A}_3} \mathbf{e}_3\, da \right) = 0, \tag{3.7}$$

or, by expanding the terms, $A_n n_2 + 0 - A_2 + 0 = 0$, so that $A_2 = A_n n_2$. We conclude that the areas of the non-slanted faces of the tetrahedron can be calculated from the area of the slanted face as

$$A_1 = A_n n_1, \qquad A_2 = A_n n_2, \qquad A_3 = A_n n_3. \tag{3.8}$$

3.2 The Theory of Stress: Cauchy's Theorem

Let us now impose the equilibrium of the region described by the tetrahedron. The balance of the forces applied on its surface reads as

$$\int_{\mathcal{A}_n} \mathbf{t}(\mathbf{n})da + \int_{\mathcal{A}_1} \mathbf{t}(-\mathbf{e}_1)da + \int_{\mathcal{A}_2} \mathbf{t}(-\mathbf{e}_2)da + \int_{\mathcal{A}_3} \mathbf{t}(-\mathbf{e}_3)da = \mathbf{0}. \tag{3.9}$$

When we shrink the tetrahedron to zero while keeping its aspect ratio unchanged, we can approximate this equation as

$$A_n \mathbf{t}(\mathbf{n}) + A_1 \mathbf{t}(-\mathbf{e}_1) + A_2 \mathbf{t}(-\mathbf{e}_2) + A_3 \mathbf{t}(-\mathbf{e}_3) = \mathbf{0}, \tag{3.10}$$

so that, by using (3.8), together with Cauchy's Lemma (3.5), we finally obtain

$$\mathbf{t}(\mathbf{n}) = \mathbf{t}(\mathbf{e}_1)n_1 + \mathbf{t}(\mathbf{e}_2)n_2 + \mathbf{t}(\mathbf{e}_3)n_3. \tag{3.11}$$

Let us call $(\sigma_{i1}, \sigma_{i2}, \sigma_{i3})$ the components of the vector $\mathbf{t}(\mathbf{e}_i)$, so that

$$\mathbf{t}(\mathbf{e}_1) = \sigma_{11}\mathbf{e}_1 + \sigma_{12}\mathbf{e}_2 + \sigma_{13}\mathbf{e}_3,$$
$$\mathbf{t}(\mathbf{e}_2) = \sigma_{21}\mathbf{e}_1 + \sigma_{22}\mathbf{e}_2 + \sigma_{23}\mathbf{e}_3,$$
$$\mathbf{t}(\mathbf{e}_3) = \sigma_{31}\mathbf{e}_1 + \sigma_{32}\mathbf{e}_2 + \sigma_{33}\mathbf{e}_3. \tag{3.12}$$

Notice how these components are independent of $\mathbf{n}$. Now we see that (3.11) can be written as the following system:

$$\mathbf{t}(\mathbf{n}) = \begin{bmatrix} \sigma_{11} & \sigma_{21} & \sigma_{31} \\ \sigma_{12} & \sigma_{22} & \sigma_{32} \\ \sigma_{13} & \sigma_{23} & \sigma_{33} \end{bmatrix} \begin{bmatrix} n_1 \\ n_2 \\ n_3 \end{bmatrix}. \tag{3.13}$$

The conclusion is that there exists a second-order tensor $\boldsymbol{\sigma}$, called the *Cauchy stress tensor*, such that for each unit vector $\mathbf{n}$, the surface traction is

$$\mathbf{t}(\mathbf{n}) = \boldsymbol{\sigma}^\mathsf{T}\mathbf{n}, \tag{3.14}$$

where $\boldsymbol{\sigma}$ is independent of $\mathbf{n}$ (but can depend on $\mathbf{x}$). This is the first point of *Cauchy's theorem*: the surface traction $\mathbf{t}(\mathbf{n})$ depends linearly on the orientation $\mathbf{n}$, which means that $\mathbf{t}(a_1\mathbf{n}_1 + a_2\mathbf{n}_2) = a_1\mathbf{t}(\mathbf{n}_1) + a_2\mathbf{t}(\mathbf{n}_2)$ for all numbers a_1, a_2 and vectors $\mathbf{n}_1, \mathbf{n}_2$.

Relying upon the stress tensor, we can write a local version of the integral equilibrium equations (3.3). With the stress tensor and the divergence theorem, we find that the resultant force over a closed region $\mathcal{R}_c$ is

$$\mathbf{f}(\mathcal{R}_c) = \int_{\partial \mathcal{R}_c} \mathbf{t}(\mathbf{n})\, da = \int_{\partial \mathcal{R}_c} \boldsymbol{\sigma}^\mathsf{T}\mathbf{n}\, da = \int_{\mathcal{R}_c} \operatorname{div} \boldsymbol{\sigma}\, dv. \tag{3.15}$$

Fig. 3.4 Balance of moments about e_1

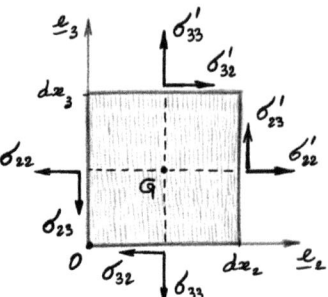

But because $\mathcal{R}_c$ is arbitrary, the integral requirement $\mathbf{f}(\mathcal{R}_c) = \mathbf{0}$ leads to the *local equilibrium equation*,

$$\text{div}\, \boldsymbol{\sigma} = \mathbf{0}, \tag{3.16}$$

which must hold everywhere in the material body.

Finally, we use Cauchy's theorem to write a local version of the balance of moments $(3.3)_2$. We now consider a cube with infinitesimal edges (dx_1, dx_2, dx_3) and we write the equilibrium of moments relative to the axis e_1, passing through the centre of mass $\mathbf{G}$ of the cube; see Fig. 3.4. The relevant stress components are not uniform throughout the cube, as depicted in the side figure. Indeed, if σ_{ij} are the stress components on the faces passing through the origin $\mathbf{O}$, then the corresponding stresses on the opposite faces are given by the Taylor expansion

$$\sigma'_{ij} = \sigma_{ij} + \sum_{k=1}^{3} \frac{\partial \sigma_{ij}}{\partial x_k} dx_k, \tag{3.17}$$

plus higher-order terms.

We now write that the moment of all forces acting on the faces of the cube is zero. In particular, with reference to the component of the moment along e_1, we have

$$\left(\sigma'_{32} + \sigma_{32}\right) dA_3 \frac{dx_3}{2} - \left(\sigma'_{23} + \sigma_{23}\right) dA_2 \frac{dx_2}{2} = 0, \tag{3.18}$$

where $dA_3 = dx_1 dx_2$ and $dA_2 = dx_1 dx_3$. Note that the normal stresses σ_{ii} play no role in this equation. Thus, bearing in mind (3.17), and neglecting higher-order terms $dx_1^2, dx_1 dx_2$, etc., the balance of moments along e_1 leads to

$$\sigma_{32} = \sigma_{23}. \tag{3.19}$$

By repeating the same reasoning for the axes e_2 and e_3, we finally conclude that *the Cauchy stress tensor is symmetric*:

$$\boldsymbol{\sigma} = \boldsymbol{\sigma}^\mathsf{T}. \tag{3.20}$$

3.3 Normal and Shear Stresses

Suppose that an element of area da on a surface S with unit normal $\mathbf{n}$ is subjected to a contact force $\mathbf{t}(\mathbf{n})da$. The *normal component* of the stress vector, denoted σ, is defined as

$$\sigma = \mathbf{t}(\mathbf{n}) \cdot \mathbf{n} = \boldsymbol{\sigma}^T \mathbf{n} \cdot \mathbf{n}. \tag{3.21}$$

This is called the *normal stress* on the surface S. It is tensile when σ is positive and compressive when σ is negative. The component of $\mathbf{t}(\mathbf{n})$ along the surface S, denoted by $\boldsymbol{\tau}$, is the *tangential stress vector* and is defined as

$$\boldsymbol{\tau} = \mathbf{t}(\mathbf{n}) - \sigma \mathbf{n}, \qquad \tau = \|\mathbf{t}(\mathbf{n}) - \sigma \mathbf{n}\|, \tag{3.22}$$

and we refer to τ as the *shear stress*; see Fig. 3.5a.

In light of the definitions above, we see that each diagonal component σ_{ii} of the stress tensor is the normal stress relative to the plane with normal $\mathbf{e}_i$, because $\sigma_{ii} = \mathbf{t}(\mathbf{e}_i) \cdot \mathbf{e}_i$. Likewise, each off-diagonal component σ_{ij} is the magnitude of the shear stress relative to plane $\mathbf{e}_i$, projected along the direction $\mathbf{e}_j$, because $\sigma_{ij} = \mathbf{t}(\mathbf{e}_i) \cdot \mathbf{e}_j$; see Fig. 3.5b.

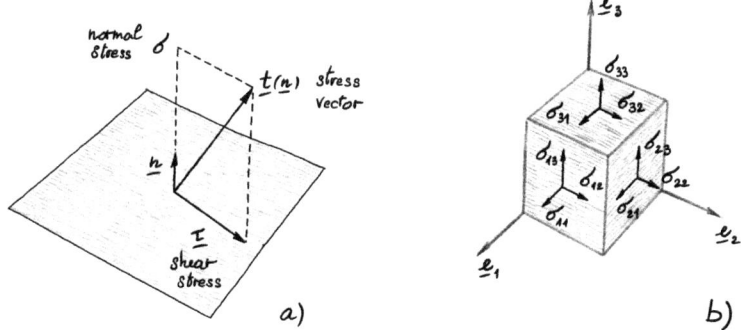

Fig. 3.5 (a) The stress vector $\mathbf{t}(\mathbf{n}) = \boldsymbol{\sigma}^T \mathbf{n}$ can be decomposed into the normal stress $\sigma = \mathbf{n} \cdot \mathbf{t}(\mathbf{n}) = \mathbf{n} \cdot (\boldsymbol{\sigma}^T \mathbf{n})$ and the tangential stress vector $\boldsymbol{\tau}$. (b) Normal and shear stresses on the coordinate faces of a cube

3.4 Principal Stresses and Principal Axes of Stress

Similarly to strain, we can find the principal stresses and principal axes of stress by solving the eigenproblem

$$\sigma \mathbf{n} = s\mathbf{n}, \tag{3.23}$$

where the s's are scalars and the $\mathbf{n}$'s are unit vectors. In practice, we solve first $\det(\sigma - s\mathbf{I}) = 0$ and find the *principal stresses* s_1, s_2, s_3, which are real numbers because σ is symmetric (see Sect. 1.7). Then we normalise the corresponding eigenvectors $(\mathbf{n}_1, \mathbf{n}_2, \mathbf{n}_3)$ to be of unit length. Because σ is symmetric, they are orthogonal, and we take them to form a direct orthonormal basis, along the directions of the *principal axes of stress*. In contrast to the eigenvalues of the strain tensors, σ is not necessarily positive definite, and the s_i may be of any sign.

Hence, the stress tensor has a diagonal representation in the coordinate system of its eigenvectors,

$$[\sigma] = \begin{bmatrix} s_1 & 0 & 0 \\ 0 & s_2 & 0 \\ 0 & 0 & s_3 \end{bmatrix}, \tag{3.24}$$

in the $(\mathbf{n}_1, \mathbf{n}_2, \mathbf{n}_3)$ basis.

3.5 Some States of Stress

Recall that the equations of equilibrium read $\operatorname{div}\sigma = \mathbf{0}$. Thus, they are automatically satisfied when σ has constant components, i.e. when the stress is *homogeneous*. We now present three of the most common states of homogeneous stress.

In Fig. 3.6, we describe three states of stress for a cube of side $2a$, with the origin located in the centre of mass.

First, consider that all six faces of the cube are under the same constant and normal force per unit area, directed inward, of magnitude p; see Fig. 3.6a. This is called a state of *uniform hydrostatic pressure* and p is called the pressure. Looking, for example, at the face at $x_1 = a$, we have $\mathbf{t}(\mathbf{e}_1) = -p\,\mathbf{e}_1$. But $\mathbf{t}(\mathbf{e}_1) = [\sigma_{11}, \sigma_{12}, \sigma_{13}]^\mathsf{T}$, showing that $\sigma_{11} = -p$ and $\sigma_{12} = \sigma_{13} = 0$, and similarly for the other faces. Hence, hydrostatic pressure corresponds to the state of stress

$$[\sigma] = \begin{bmatrix} -p & 0 & 0 \\ 0 & -p & 0 \\ 0 & 0 & -p \end{bmatrix} = -p\mathbf{I}. \tag{3.25}$$

3.6 The Nominal Stress Tensor

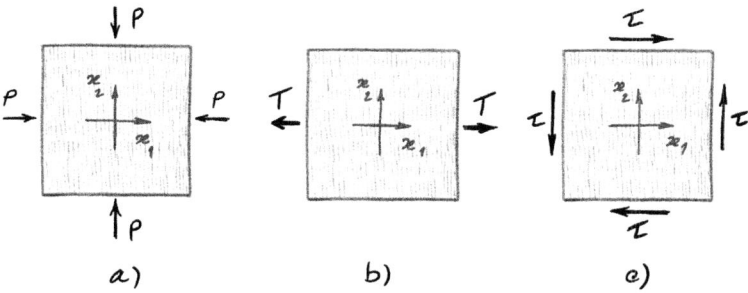

Fig. 3.6 Uniform states of stress: (**a**) hydrostatic pressure, (**b**) uni-axial stress, and (**c**) shear stress. For simplicity, we only show the face perpendicular to $\mathbf{e}_3$.

Next, we consider that the cube is put under constant tension or compression T (force per area unit) along x_1; see Fig. 3.6b. Looking at the face at $x_1 = a$, we have $\mathbf{t}(\mathbf{e}_1) = T\mathbf{e}_1$. But $\mathbf{t}(\mathbf{e}_1) = [\sigma_{11}, \sigma_{12}, \sigma_{13}]^T$, showing that $\sigma_{11} = T$ and $\sigma_{12} = \sigma_{13} = 0$. Looking at the faces at $x_2 = a$, $x_3 = a$, we find that the other components of σ are zero, so that

$$[\sigma] = \begin{bmatrix} T & 0 & 0 \\ 0 & 0 & 0 \\ 0 & 0 & 0 \end{bmatrix}, \qquad (3.26)$$

in the $(\mathbf{e}_1, \mathbf{e}_2, \mathbf{e}_3)$ basis. We say that the body is under *uni-axial stress*: $T > 0$ corresponds to uni-axial tension and $T < 0$ corresponds to uni-axial compression.

Finally, consider that tangential, constant forces per unit area τ are applied as depicted in Fig. 3.6c. Here, the traction on the face $x_1 = a$ is $\mathbf{t}(\mathbf{e}_1) = \tau\mathbf{e}_2$, but $\mathbf{t}(\mathbf{e}_1) = [\sigma_{11}, \sigma_{12}, \sigma_{13}]^T$, showing that $\sigma_{12} = \tau$ and $\sigma_{11} = \sigma_{13} = 0$. Likewise, the traction on the face $x_2 = a$ is $\mathbf{t}(\mathbf{e}_2) = \tau\mathbf{e}_1$ so, because $\mathbf{t}(\mathbf{e}_2) = [\sigma_{21}, \sigma_{22}, \sigma_{23}]^T$, we find $\sigma_{21} = \tau$ and $\sigma_{22} = \sigma_{23} = 0$. On the face $x_3 = a$, there are no forces applied, leading to $\sigma_{31} = \sigma_{32} = \sigma_{33} = 0$, so that finally

$$[\sigma] = \begin{bmatrix} 0 & \tau & 0 \\ \tau & 0 & 0 \\ 0 & 0 & 0 \end{bmatrix}, \qquad (3.27)$$

in the $(\mathbf{e}_1, \mathbf{e}_2, \mathbf{e}_3)$ basis. We say that the body is in a state of *uniform shear stress*.

3.6 The Nominal Stress Tensor

During a tensile test, a specimen is clamped and stretched; see Fig. 3.7. In general, the stretch is measured in real time throughout the deformation, and so is the force,

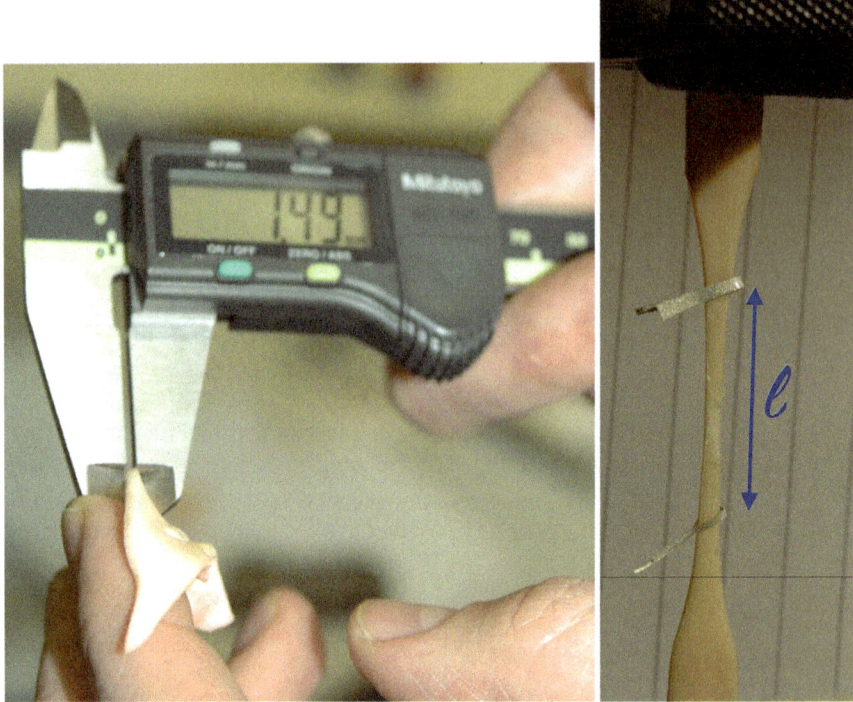

Fig. 3.7 Left: Measuring the *reference* cross-sectional dimensions of a sample with a Vernier calliper. Right: Clamping the sample in a tensile machine and tracking the *current* length ℓ using two reflective markers, with positions tracked by a vertical laser beam

by using a load cell. To compute the Cauchy stress, one would need to measure the cross-sectional area of the specimen in real time as well, which is no easy task. A more practical way to compute the stress is to link it to the undeformed cross-sectional area of the specimen, leading to the introduction of the *nominal stress tensor*, as follows.

Using Nanson's formula (2.25), the traction on an area element $\mathbf{n}da$ in the current configuration is

$$d\mathbf{f} = \mathbf{t(n)}da = \boldsymbol{\sigma}^\mathsf{T}\mathbf{n}da = J\boldsymbol{\sigma}^\mathsf{T}(\mathsf{F}^{-1})^\mathsf{T}\mathbf{N}dA \tag{3.28}$$

which can be rewritten as

$$d\mathbf{f} = \mathsf{S}^\mathsf{T}\mathbf{N}dA \qquad \text{where} \qquad \mathsf{S} := J\mathsf{F}^{-1}\boldsymbol{\sigma} \tag{3.29}$$

is the *nominal stress tensor*, also referred to as the *engineering stress*. Its transpose, $\mathsf{P} = \mathsf{S}^\mathsf{T} = J\boldsymbol{\sigma}(\mathsf{F}^{-1})^\mathsf{T}$, is the so-called *first Piola-Kirchhoff stress tensor*.

3.7 Boundary Conditions

The nominal stress S measures the force *per unit reference area*, and the Cauchy stress $\boldsymbol{\sigma}$ measures the force *per unit deformed area*. Of course, $\boldsymbol{\sigma}$ and S are connected through (3.29), so ultimately it is a matter of choice to prefer recording one measure of stress over the other.

The first equilibrium equation in the absence of body forces, div $\boldsymbol{\sigma} = \mathbf{0}$, can be recast in terms of the nominal stress S as follows: First, write the integral form of the balance equation, i.e.

$$\mathbf{f}(\mathcal{R}_c) = \int_{\partial \mathcal{R}_c} \boldsymbol{\sigma}^\mathsf{T} \mathbf{n} \, da = \mathbf{0}. \tag{3.30}$$

Then convert the integral over the current configuration to an integral over the reference configuration using Nanson's formula, leading to

$$\int_{\partial \mathcal{R}_r} \mathsf{S}^\mathsf{T} \mathbf{N} \, dA = \mathbf{0}, \tag{3.31}$$

and finally, by the divergence theorem, to

$$\text{Div } \mathsf{S} = \mathbf{0}. \tag{3.32}$$

The second equilibrium equation, $\boldsymbol{\sigma} = \boldsymbol{\sigma}^\mathsf{T}$, implies that $\mathsf{FS} = \mathsf{S}^\mathsf{T}\mathsf{F}^\mathsf{T}$, a condition that represents the balance of angular momentum in terms of the nominal stress. It also shows that $\mathsf{S} = \mathsf{F}^{-1}\mathsf{S}^\mathsf{T}\mathsf{F}^\mathsf{T}$, which is different from S^T in general. We conclude that S is generally not symmetric.

We have established two different ways of writing the equations of equilibrium: a Eulerian formulation in terms of $\boldsymbol{\sigma}$ and a Lagrangian formulation in terms of S. Which one is the most convenient to use depends on the problem at hand.

3.7 Boundary Conditions

Generally, solving a problem of nonlinear elasticity means solving the equations of equilibrium for a given body, subjected to some boundary conditions prescribed on its surface, typically displacements and/or tractions.

In a Lagrangian formulation, boundary tractions are written using the nominal stress tensor

$$\mathsf{S}^\mathsf{T}\mathbf{N} = \mathbf{s} \quad \text{on} \quad \partial \mathcal{R}_r, \tag{3.33}$$

where the vector **s**, representing forces applied per unit area on the boundary in the reference configuration, is prescribed.

When **s** depends on the point of application only, $\mathbf{s} = \mathbf{s}(\mathbf{X})$, we talk of a "dead load", as it is independent of the deformation of the body (hence, independent of

F). Some loads, however, may depend on the deformation. For example, take a solid subjected to the hydrostatic pressure $\boldsymbol{\sigma} = -p\mathbf{I}$ on $\partial\mathcal{R}_c$. According to (3.29), it corresponds to surface traction $\mathbf{S}^T\mathbf{N} = -pJ\mathbf{F}^{-T}\mathbf{N}$ on $\partial\mathcal{R}_r$, and therefore $\mathbf{s} = -pJ(\mathbf{F}^{-1})^T\mathbf{N}$ depends on the deformation.

3.8 Work

We compute the work done by the forces (surface tractions only here) when the body is deformed slightly in the current configuration, i.e. when a point at position $\mathbf{x}$ is displaced to $\mathbf{x}' = \mathbf{x} + \boldsymbol{\delta}\mathbf{x}$, where $\boldsymbol{\delta}\mathbf{x}$ is an infinitesimal displacement. The work is

$$\text{Work} = \int_{\partial\mathcal{R}_c} \mathbf{t}(\mathbf{n}) \cdot \boldsymbol{\delta}\mathbf{x} \, da = \int_{\partial\mathcal{R}_r} \mathbf{S}^T\mathbf{N} \cdot \boldsymbol{\delta}\mathbf{x} \, dA = \int_{\partial\mathcal{R}_r} (\mathbf{S}\boldsymbol{\delta}\mathbf{x}) \cdot \mathbf{N} \, dA$$

$$= \int_{\mathcal{R}_r} \text{Div}(\mathbf{S}\boldsymbol{\delta}\mathbf{x}) dV = \int_{\mathcal{R}_r} [(\text{Div}\,\mathbf{S}) \cdot \boldsymbol{\delta}\mathbf{x} + \text{tr}\,(\mathbf{S}\,\text{Grad}\,\boldsymbol{\delta}\mathbf{x})] dV, \quad (3.34)$$

where we used, in turn, the formula $\mathbf{u} \cdot \mathbf{T}\mathbf{v} = \mathbf{v} \cdot \mathbf{T}^T\mathbf{u}$, the divergence theorem, and the identity (left as an exercise) $\text{Div}\,(\mathbf{T}\mathbf{u}) = \mathbf{u} \cdot \text{Div}\,\mathbf{T} + \text{tr}\,(\mathbf{T}\,\text{Grad}\,\mathbf{u})$ for vectors $\mathbf{u}, \mathbf{v}$ and tensors $\mathbf{T}$.

At this point, because $\text{Div}\,\mathbf{S} = \mathbf{0}$ by the equation of equilibrium (3.32), and because $\text{Grad}\,\boldsymbol{\delta}\mathbf{x} = \text{Grad}\,\mathbf{x}' - \text{Grad}\,\mathbf{x} = \mathbf{F}' - \mathbf{F} = \boldsymbol{\delta}\mathbf{F}$, the infinitesimal change in the deformation gradient $\mathbf{F}$, we can conclude that

$$\text{Work} = \int_{\mathcal{R}_r} \text{tr}\,(\mathbf{S}\boldsymbol{\delta}\mathbf{F}) dV. \quad (3.35)$$

Chapter 4
Constitutive Equations

In the previous chapters, we discussed deformation and stress separately. In this chapter, we look at how these fields are connected (as they should be, because clearly, either one leads to the other). We arrive at a three-dimensional (3D) generalisation of *Hooke's law*, encountered in the early study of mechanics, connecting the elongation of a one-dimensional (1D) spring to the elastic force. Here the connection between the deformation gradient and the stress tensor is made thanks to the *principle of work*.

4.1 The Principle of Work

We first present the principle of work in the 1D case. The pillar of the theory of elasticity is *Hooke's law*, stating that the force F required to produce an elongation x in a one-dimensional elastic spring with stiffness k is

$$F = kx. \tag{4.1}$$

The work done by the applied force to further stretch the spring by an infinitesimal amount δx (here the delta stands for "small variation") is

$$\text{Work} = F\delta x = (kx)\delta x. \tag{4.2}$$

Calling $U = \frac{1}{2}kx^2$ the "potential elastic energy" of the spring and neglecting the $(\delta x)^2$ term, we find that Work $= \delta U$, which leads to the statement of the *principle of work*: the work done to stretch a spring equals the change of elastic potential energy. The term "potential" here signifies that the work depends only on the difference in elastic energy between an initial state (defined by x) and a final state (defined by $x + \delta x$), and not on the path between the two states.

The application of the principle of work in 1D led to the characterisation of the potential elastic energy. We now follow the same procedure in 3D, with the difference that stress replaces force, and deformation replaces elongation.

Consider a 3D body occupying the region $\mathcal{R}_r$ in the reference configuration $\mathcal{B}_r$. If the only mechanical forces in place are surface tractions, we saw at the end of the previous chapter that their work for a small increment of deformation gradient is

$$\text{Work} = \int_{\mathcal{R}_r} \text{tr}\,(\mathsf{S}\,\delta\mathsf{F}) dV. \tag{4.3}$$

The statement of the principle of work seen in 1D holds in 3D as well: we can manipulate the right-hand side of (4.3) to recast it as the difference in potential energy between an initial and a final state,

$$\int_{\mathcal{R}_r} \text{tr}\,(\mathsf{S}\,\delta\mathsf{F}) dV = \delta U, \tag{4.4}$$

where U is the elastic potential energy stored in the body. We may now introduce a local scalar function W, the *stored energy density*, computed per unit volume in the reference configuration, such that $U = \int_{\mathcal{R}_r} W\,dV$. Then, because the coordinates and the volumes of the reference configuration are fixed,

$$\delta U = \delta \left(\int_{\mathcal{R}_r} W\,dV \right) = \int_{\mathcal{R}_r} \delta(W\,dV) = \int_{\mathcal{R}_r} (\delta W) dV. \tag{4.5}$$

Comparing (4.4) and (4.5), because the region $\mathcal{R}_r$ is arbitrary, it follows that locally for each volume element $\delta W = \text{tr}\,(\mathsf{S}\,\delta\mathsf{F})$.

Because the work done and the change in stored energy must be path independent, we may then write this equality for the full differentials, as

$$dW = \text{tr}\,(\mathsf{S}\,d\mathsf{F}). \tag{4.6}$$

We now assume that the stored energy density is a function of the deformation gradient only: $W = W(\mathsf{F})$, placing ourselves in the framework of what is called *Green elasticity*, or *hyperelasticity* (other theories assume further dependencies, for example, that W also depends on gradients of F). Then the previous equation reads

$$\sum_{i=1}^{3} \sum_{j=1}^{3} \frac{\partial W}{\partial F_{ij}} dF_{ij} = \sum_{i=1}^{3} \sum_{j=1}^{3} S_{ji} dF_{ij}. \tag{4.7}$$

These expressions must coincide for all differentials $d\mathsf{F}$, leading to the conclusion that

$$\mathsf{S} = \frac{\partial W}{\partial \mathsf{F}}, \qquad S_{ji} = \frac{\partial W}{\partial F_{ij}}, \tag{4.8}$$

where $\partial W/\partial F$ is the second-order tensor with components defined by the convention

$$\left(\frac{\partial W}{\partial F}\right)_{ji} = \frac{\partial W}{\partial F_{ij}}. \tag{4.9}$$

By recalling the connection $S = JF^{-1}\sigma$ between the nominal stress S and the Cauchy stress σ, we obtain

$$\sigma = J^{-1}F\frac{\partial W}{\partial F}, \quad \sigma_{ij} = J^{-1}\sum_{k=1}^{3} F_{ik}\frac{\partial W}{\partial F_{jk}}, \tag{4.10}$$

which provides a formula for σ in terms of $W(F)$.

Once W is prescribed for a given material, we may impose the deformation, compute F, and then compute the stress through the formulas above, and verify that the equation of equilibrium is satisfied.

We next see that $W(F)$ cannot be prescribed completely arbitrarily and must accommodate some constraints imposed by Physics.

4.2 Objectivity

What happens to physical quantities when a body is first deformed and subsequently translated and rotated?

Consider a body which is first subjected to a deformation from $\mathcal{B}_r$ (coordinates X) to $\mathcal{B}_c$ (coordinates x) through $x = \chi(X)$, and then to a translation and rotation from $\mathcal{B}_c$ to $\mathcal{B}_c^*$ (coordinates x^*); see Fig. 4.1. Mathematically, we write

$$x^* = c + Qx, \tag{4.11}$$

where Q is a proper orthogonal tensor and c is a vector. We call χ^* the deformation bringing a point at X in $\mathcal{B}_r$ to x^* in $\mathcal{B}_c^*$, with gradient $\partial \chi^*/\partial x = Q$.

First, we find that the distance between any two points in the solid, one at x and the other at y, is not affected by the roto-translation, because $x^* - y^* = Qx - Qy = Q(x - y)$, so that

$$|x^* - y^*|^2 = (x^* - y^*) \cdot (x^* - y^*) = Q(x - y) \cdot Q(x - y)$$
$$= (x - y) \cdot Q^TQ(x - y) = |x - y|^2. \tag{4.12}$$

Fig. 4.1 An objective quantity is unaffected by a superposed roto-translation **q** onto a deformation χ

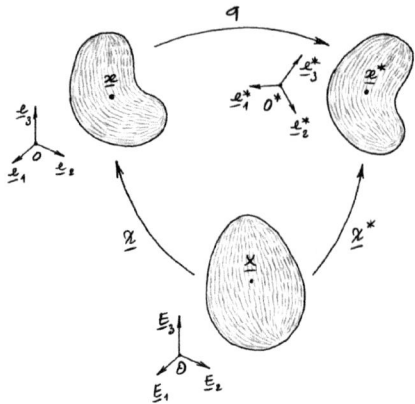

We then look at the total deformation gradient

$$\mathsf{F}^* := \frac{\partial \chi}{\partial \mathbf{x}} = \mathsf{QF}, \tag{4.13}$$

which is the composition of the pre-deformation **F**, followed by the rotation **Q** (see Sect. 2.6).

In Mechanics, certain properties and quantities are "objective", which means that they are unaffected by roto-translations. For example, we take it for granted that distances (as we just saw), temperature, or mass are such quantities. It is an axiom of Mechanics that *the material response is objective*. It means that the stored energy density W must be such that

$$W(\mathsf{F}) = W(\mathsf{F}^*), \qquad \mathsf{F}^* = \mathsf{QF}, \tag{4.14}$$

for all rotations **Q** and deformation gradients **F**.

An important consequence of the axiom of objectivity is that the stored energy does not depend on the whole tensor **F**. In fact, by the polar decomposition of **F**, we have $W(\mathsf{F}) = W(\mathsf{QRU})$. In particular, objectivity must hold when $\mathsf{Q} = \mathsf{R}^\mathsf{T}$, which gives

$$W = W(\mathsf{U}), \tag{4.15}$$

and, by the square root theorem,

$$W = W(\mathsf{C}). \tag{4.16}$$

In summary, the axiom of objectivity reveals that the stored energy density, which we thought was a function of nine scalars (the components of **F**), is actually a function of six scalars (the independent components of the symmetric tensor **C**).

4.3 Material Symmetry and Isotropic Hyperelasticity

Some materials do not behave in the same way in different directions. For example, when a solid is reinforced with parallel fibres, it has a much stiffer response to forces along their common direction than in the plane perpendicular to that direction.

Consider the experiment described in Fig. 4.2. There, a cube made of a soft matrix reinforced by stiff fibres along $\mathbf{E}_3$ in the reference configuration $\mathcal{B}_r$ is stretched by a deformation F, leading to the current configuration $\mathcal{B}_c$. If prior to the application of the deformation F, the cube is rotated around $\mathbf{E}_3$ by a rotation P_3, then the overall deformation $\mathsf{F}\mathsf{P}_3$ leads the cube to a new current configuration $\mathcal{B}'_c$. But because $\mathcal{B}_c$ and $\mathcal{B}'_c$ are indistinguishable, we expect that the response of the body is unaffected by pre-imposed rotation P_3, which is then called a *symmetry transformation*.

On the other hand, in the experiment described in Fig. 4.3, the pre-imposed rotation P_1 around $\mathbf{E}_1$ (which is perpendicular to the direction of reinforcement) leads to different current configurations $\mathcal{B}_c$ and $\mathcal{B}''_c$ and different responses when stretched, so that P_1 is not a symmetry transformation in our example.

Formally, we say that P is a symmetry transformation if

$$W(\mathsf{F}) = W(\mathsf{F}\mathsf{P}), \tag{4.17}$$

for all possible deformations F. So, for instance, we can expect that $W(\mathsf{F}) = W(\mathsf{F}\mathsf{P}_3)$ for all F in our example. Notice that while with objectivity we were asking that the energy is unaffected by *superposed* rotations, in the case of symmetry transformations, we ask that the energy is unaffected by *pre-imposed* rotations.

A material is said to be *isotropic* when its mechanical behaviour is unaffected by *any rotation* P taking place prior to a given deformation. Physically, it means that the response of a small specimen of material is independent of its orientation. Then, for isotropic materials, Eq. (4.17) must hold for any deformation F and rotation P.

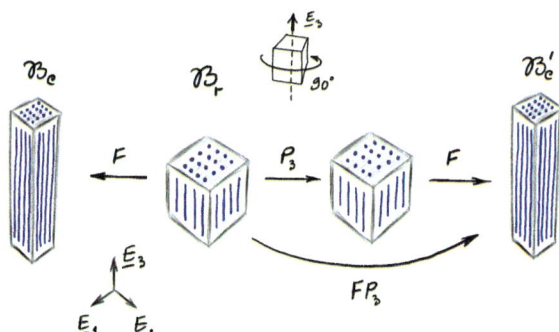

Fig. 4.2 The pre-imposed rotation P_3 is a symmetry transformation

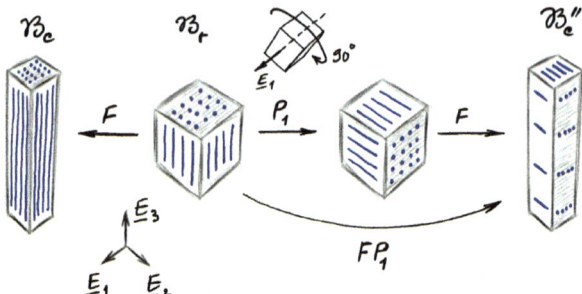

Fig. 4.3 The pre-imposed rotation P_1 is *not* a symmetry transformation

Invoking objectivity also, we may thus write for $W = W(C)$ that $W(F^T F) = W((FP)^T(FP)) = W(P^T F^T FP)$, or

$$W(C) = W(P^T CP), \qquad \text{for all rotations P.} \tag{4.18}$$

A remarkable consequence of isotropy is that the energy does not depend on six independent components of the tensor C, but only on its three eigenvalues, as we now see. First we note that C and $P^T CP$ have the same eigenvalues, because

$$\det(P^T CP - \lambda^2 I) = \det(P^T CP - \lambda^2 P^T P)$$
$$= \det[P^T (C - \lambda^2 I) P] = \det[PP^T (C - \lambda^2 I)] = \det(C - \lambda^2 I). \tag{4.19}$$

And then, we realise that C and $P^T CP$ do not have the same eigenvectors, because, for a general rotation, $P^T CP\mathbf{u} \neq \lambda^2 \mathbf{u}$ when $C\mathbf{u} = \lambda^2 \mathbf{u}$. Hence, the only elements in common between C and the whole family of tensors $P^T CP$ spanning all rotations P are the three eigenvalues $\lambda_1^2, \lambda_2^2, \lambda_3^2$. Then, enforcing (4.18) means that W depends on $\lambda_1^2, \lambda_2^2, \lambda_3^2$ at most. In other words, isotropy restricts the way the energy depends on C:

$$W(C) = W(\lambda_1, \lambda_2, \lambda_3). \tag{4.20}$$

In $(\mathbf{u}^{(1)}, \mathbf{u}^{(2)}, \mathbf{u}^{(3)})$, the orthonormal basis of the eigenvectors of C, the right Cauchy-Green strain tensor, is written as

$$[C] = \begin{bmatrix} \lambda_1^2 & 0 & 0 \\ 0 & \lambda_2^2 & 0 \\ 0 & 0 & \lambda_3^2 \end{bmatrix}. \tag{4.21}$$

4.3 Material Symmetry and Isotropic Hyperelasticity

Now take the particular rotation about $\mathbf{u}^{(3)}$ such that

$$\mathbf{P}\mathbf{u}^{(1)} = -\mathbf{u}^{(2)}, \qquad \mathbf{P}\mathbf{u}^{(2)} = \mathbf{u}^{(1)}, \qquad \mathbf{P}\mathbf{u}^{(3)} = \mathbf{u}^{(3)}. \tag{4.22}$$

Then, in the eigenvector orthonormal basis,

$$[\mathbf{P}^\mathsf{T}\mathbf{C}\mathbf{P}] = \begin{bmatrix} 0 & -1 & 0 \\ 1 & 0 & 0 \\ 0 & 0 & 1 \end{bmatrix} \begin{bmatrix} \lambda_1^2 & 0 & 0 \\ 0 & \lambda_2^2 & 0 \\ 0 & 0 & \lambda_3^2 \end{bmatrix} \begin{bmatrix} 0 & 1 & 0 \\ -1 & 0 & 0 \\ 0 & 0 & 1 \end{bmatrix} = \begin{bmatrix} \lambda_2^2 & 0 & 0 \\ 0 & \lambda_1^2 & 0 \\ 0 & 0 & \lambda_3^2 \end{bmatrix}, \tag{4.23}$$

that is, λ_1^2 and λ_2^2 have swapped positions. It follows that $W(\lambda_1, \lambda_2, \lambda_3) = W(\lambda_2, \lambda_1, \lambda_3)$, and we may proceed similarly with other rotations about the eigenvectors. Therefore, we have learned that isotropy implies that the order of λ_i is irrelevant, so W must be a *symmetric function* of $\lambda_1, \lambda_2, \lambda_3$:

$$W(\lambda_1, \lambda_2, \lambda_3) = W(\lambda_2, \lambda_1, \lambda_3) = W(\lambda_1, \lambda_3, \lambda_2) = W(\lambda_3, \lambda_2, \lambda_1). \tag{4.24}$$

As an example, consider the energy $W = c(\lambda_1^2 + \lambda_2\lambda_3 + \lambda_3^2)$, where c is a constant; clearly, swapping λ_1 with λ_2 gives a different W in general: this energy is not a symmetric function of $\lambda_1, \lambda_2, \lambda_3$ and is not appropriate for isotropic elasticity. On the other hand, $W = c(\lambda_1^2 + \lambda_2^2 + \lambda_3^2)$ is unaffected by swaps of the λ_i.

Now recall from Sect. 1.7 that $\lambda_1^2, \lambda_2^2, \lambda_3^2$ are the (real) roots of the cubic

$$\lambda^6 - I_1\lambda^4 + I_2\lambda^2 - I_3 = 0, \tag{4.25}$$

where

$$I_1 = \operatorname{tr}\mathbf{C}, \qquad I_2 = \tfrac{1}{2}\left[(\operatorname{tr}\mathbf{C})^2 - \operatorname{tr}\left(\mathbf{C}^2\right)\right], \qquad I_3 = \det\mathbf{C}, \tag{4.26}$$

are the principal invariants of $\mathbf{C}$ (and of $\mathbf{B}$), related to the λ_i through

$$I_1 = \lambda_1^2 + \lambda_2^2 + \lambda_3^2, \qquad I_2 = \lambda_1^2\lambda_2^2 + \lambda_2^2\lambda_3^2 + \lambda_3^2\lambda_1^2, \qquad I_3 = \lambda_1^2\lambda_2^2\lambda_3^2. \tag{4.27}$$

By inverting these three equations, we can (in principle) express λ_i as functions of the I_i.

In conclusion, due to the requirements of objectivity and isotropy, the energy shall depend only on the three invariants of $\mathbf{C}$,

$$W(\mathbf{C}) = W(I_1, I_2, I_3). \tag{4.28}$$

4.4 Stress-Deformation Relations in Terms of Invariants

For a hyperelastic material, the nominal stress $\mathbf{S}$ is related to W through $S_{ij} = \partial W/\partial F_{ji}$. When the material is isotropic, we have, by the chain rule,

$$S_{ij} = \frac{\partial W}{\partial I_1}\frac{\partial I_1}{\partial F_{ji}} + \frac{\partial W}{\partial I_2}\frac{\partial I_2}{\partial F_{ji}} + \frac{\partial W}{\partial I_3}\frac{\partial I_3}{\partial F_{ji}}. \tag{4.29}$$

We then compute the derivatives of the principal invariants with respect to $\mathbf{F}$; we find that

$$\frac{\partial I_1}{\partial \mathbf{F}} = 2\mathbf{F}^\mathsf{T}, \qquad \frac{\partial I_2}{\partial \mathbf{F}} = 2I_1\mathbf{F}^\mathsf{T} - 2\mathbf{F}^\mathsf{T}\mathbf{F}\mathbf{F}^\mathsf{T}, \qquad \frac{\partial I_3}{\partial \mathbf{F}} = 2I_3\mathbf{F}^{-1}. \tag{4.30}$$

To prove the first of these identities, recall that $I_1 = \sum_{k=1}^{3} C_{kk} = \sum_{k=1}^{3}\sum_{\ell=1}^{3} F_{\ell k} F_{\ell k}$, so that

$$\frac{\partial I_1}{\partial F_{ji}} = 2\sum_{k=1}^{3}\sum_{\ell=1}^{3} \frac{\partial F_{\ell k}}{\partial F_{ji}} F_{\ell k} = 2\sum_{k=1}^{3}\sum_{\ell=1}^{3} \delta_{j\ell}\delta_{ik} F_{\ell k} = 2F_{ji}. \tag{4.31}$$

Then the proof for the second identity proceeds similarly. Finally, the proof for the third identity relies on Jacobi's identity (see Sect. 1.8), which reads

$$\frac{\partial}{\partial \tau}(\det \mathbf{F}) = (\det \mathbf{F})\,\mathrm{tr}\left(\mathbf{F}^{-1}\frac{\partial \mathbf{F}}{\partial \tau}\right), \tag{4.32}$$

for any scalar variable τ. Here we write

$$\frac{\partial I_3}{\partial F_{ji}} = \frac{\partial}{\partial F_{ji}}(\det \mathbf{F})^2 = 2(\det \mathbf{F})\frac{\partial}{\partial F_{ji}}(\det \mathbf{F}) = 2(\det \mathbf{F})^2 \mathrm{tr}\left(\mathbf{F}^{-1}\frac{\partial \mathbf{F}}{\partial F_{ji}}\right),$$

$$= 2I_3 \sum_{k=1}^{3}\sum_{\ell=1}^{3} F_{k\ell}^{-1} \frac{\partial F_{\ell k}}{\partial F_{ji}} = 2I_3 \sum_{k=1}^{3}\sum_{\ell=1}^{3} F_{k\ell}^{-1}\delta_{j\ell}\delta_{ki} = 2I_3 F_{ij}^{-1}. \tag{4.33}$$

Going back now to (4.29), we thus have

$$\mathbf{S} = 2W_1 \mathbf{F}^\mathsf{T} + 2W_2(I_1\mathbf{F}^\mathsf{T} - \mathbf{F}^\mathsf{T}\mathbf{F}\mathbf{F}^\mathsf{T}) + 2I_3 W_3 \mathbf{F}^{-1}, \tag{4.34}$$

where we introduced the short-hand notation

$$W_1 = \frac{\partial W}{\partial I_1}, \qquad W_2 = \frac{\partial W}{\partial I_2}, \qquad W_3 = \frac{\partial W}{\partial I_3}. \tag{4.35}$$

4.5 Stress-Deformation Relations in Terms of Stretches

The corresponding expression for the Cauchy stress is then found from the connection $\boldsymbol{\sigma} = J^{-1}\mathbf{F}\mathbf{S}$ with $J = I_3^{1/2}$, as

$$\boldsymbol{\sigma} = 2I_3^{1/2}W_3\mathbf{I} + 2I_3^{-1/2}(W_1 + I_1 W_2)\mathbf{B} - 2I_3^{-1/2}W_2\mathbf{B}^2, \quad (4.36)$$

where $\mathbf{B} = \mathbf{F}\mathbf{F}^\mathsf{T}$ is the right Cauchy-Green deformation tensor.

An alternative representation is obtained by writing down the Cayley-Hamilton theorem for $\mathbf{B}$, see Eq. (1.36). By multiplying the identity $\mathbf{B}^3 - I_1\mathbf{B}^2 + I_2\mathbf{B} - I_3\mathbf{I} = 0$ across by $\mathbf{B}^{-1}$, we may express $\mathbf{B}^2$ in terms of $\mathbf{B}$, $\mathbf{I}$, and $\mathbf{B}^{-1}$ as $\mathbf{B}^2 = I_1\mathbf{B} - I_2\mathbf{I} + I_3\mathbf{B}^{-1}$. Then, a stress-deformation relation equivalent to (4.36) is

$$\boldsymbol{\sigma} = 2I_3^{-1/2}(I_2 W_2 + I_3 W_3)\mathbf{I} + 2I_3^{-1/2}W_1\mathbf{B} - 2I_3^{1/2}W_2\mathbf{B}^{-1}. \quad (4.37)$$

We have thus established two equivalent forms relating the Cauchy stress to the Cauchy-Green tensor:

$$\boldsymbol{\sigma} = \beta_0\mathbf{I} + \beta_1\mathbf{B} + \beta_2\mathbf{B}^2, \quad (4.38)$$

where

$$\beta_0 = 2I_3^{1/2}W_3, \qquad \beta_1 = 2I_3^{-1/2}(W_1 + I_1 W_2), \qquad \beta_2 = -2I_3^{-1/2}W_2, \quad (4.39)$$

and

$$\boldsymbol{\sigma} = \chi_0\mathbf{I} + \chi_1\mathbf{B} + \chi_{-1}\mathbf{B}^{-1}, \quad (4.40)$$

where

$$\chi_0 = 2I_3^{-1/2}(I_2 W_2 + I_3 W_3), \qquad \chi_1 = 2I_3^{-1/2}W_1, \qquad \chi_{-1} = -2I_3^{1/2}W_2. \quad (4.41)$$

4.5 Stress-Deformation Relations in Terms of Stretches

Instead of using the principal invariants I_1, I_2, I_3 as independent measures of deformation, we can use, equivalently, the stretches λ_1, λ_2, λ_3. They are connected by

$$\begin{aligned}
I_1 &= \operatorname{tr}\mathbf{C} = \lambda_1^2 + \lambda_2^2 + \lambda_3^2, \\
I_2 &= \tfrac{1}{2}[I_1^2 - \operatorname{tr}(\mathbf{C}^2)] = \lambda_1^2\lambda_2^2 + \lambda_2^2\lambda_3^2 + \lambda_3^2\lambda_1^2, \\
I_3 &= \det\mathbf{C} = \lambda_1^2\lambda_2^2\lambda_3^2 = J^2,
\end{aligned} \quad (4.42)$$

and we note that these are symmetric functions of the stretches.

In the coordinate system ($\mathbf{v}^{(1)}$, $\mathbf{v}^{(2)}$, $\mathbf{v}^{(3)}$) of the unit eigenvectors of $\mathbf{B}$, we have

$$[\mathbf{I}] = \begin{bmatrix} 1 & 0 & 0 \\ 0 & 1 & 0 \\ 0 & 0 & 1 \end{bmatrix}, \quad [\mathbf{B}] = \begin{bmatrix} \lambda_1^2 & 0 & 0 \\ 0 & \lambda_2^2 & 0 \\ 0 & 0 & \lambda_3^2 \end{bmatrix}, \quad [\mathbf{B}^2] = \begin{bmatrix} \lambda_1^4 & 0 & 0 \\ 0 & \lambda_2^4 & 0 \\ 0 & 0 & \lambda_3^4 \end{bmatrix}. \tag{4.43}$$

Because these matrices are diagonal in this basis, the Cauchy stress is also diagonal in this basis. Then, by using (4.38), we find

$$\sigma_{11} = \beta_0 + \beta_1 \lambda_1^2 + \beta_2 \lambda_1^4 = 2J^{-1}[I_3 W_3 + (W_1 + I_1 W_2)\lambda_1^2 - W_2 \lambda_1^4], \tag{4.44}$$

and similarly for σ_{22} and σ_{33}. Compare the expression for σ_{11} above with that obtained when computing $J^{-1}\lambda_1 \partial W/\partial \lambda_1$:

$$J^{-1}\lambda_1 \frac{\partial W}{\partial \lambda_1} = J^{-1}\lambda_1 \left[\frac{\partial I_1}{\partial \lambda_1} W_1 + \frac{\partial I_2}{\partial \lambda_1} W_2 + \frac{\partial I_3}{\partial \lambda_1} W_3 \right]$$

$$= J^{-1}\lambda_1 \left[2\lambda_1 W_1 + 2\lambda_1(\lambda_2^2 + \lambda_3^2) W_2 + 2\lambda_1 \lambda_2^2 \lambda_3^2 W_3 \right]$$

$$= 2J^{-1} \left[\lambda_1^2 W_1 + (\lambda_1^2 \lambda_2^2 + \lambda_1^2 \lambda_3^2) W_2 + I_3 W_3 \right]. \tag{4.45}$$

These expressions are indeed the same. We conclude that the principal Cauchy stresses are given by

$$\sigma_1 = J^{-1}\lambda_1 \frac{\partial W}{\partial \lambda_1}, \qquad \sigma_2 = J^{-1}\lambda_2 \frac{\partial W}{\partial \lambda_2}, \qquad \sigma_3 = J^{-1}\lambda_3 \frac{\partial W}{\partial \lambda_3}. \tag{4.46}$$

4.6 Incompressible Hyperelastic Solids

An *incompressible* material can only accommodate isochoric deformations. It follows that the deformation gradient $\mathbf{F}$ must satisfy the *internal constraint*

$$J = \det \mathbf{F} = 1, \tag{4.47}$$

at each point of the material.

Mathematically speaking, "$g(\mathbf{F}) = \det \mathbf{F} - 1 = 0$" is a constraint. What does "enforcing a constraint" mean physically? Consider the example in Fig. 4.4a of a spring with stiffness k attached at end point A, with a force $\mathbf{f}$ applied at the other end point B. Here B may be displaced anywhere in space, as the spring is unconstrained. Equilibrium means that $\mathbf{f} - k\mathbf{x} = \mathbf{0}$, and the work done by the force with a small displacement $\mathbf{dx}$ is $\mathbf{f} \cdot \mathbf{dx} = k\mathbf{x} \cdot \mathbf{dx}$, which is the vector counterpart of (4.2). In

4.6 Incompressible Hyperelastic Solids

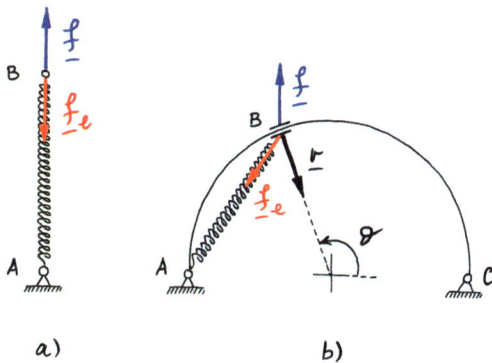

Fig. 4.4 Unconstrained system (**a**) and constrained system (**b**)

Fig. 4.4b, the end point B is *constrained* to stay on the wire. Equilibrium is now $\mathbf{f} - k\mathbf{x} - \mathbf{r} = \mathbf{0}$, where $\mathbf{r}$ is the (normal) reaction of the smooth wire. But the work done by the force is still $\mathbf{f} \cdot \mathbf{dx} = k\mathbf{x} \cdot \mathbf{dx}$, because $\mathbf{r} \cdot \mathbf{dx} = 0$: the reaction force is workless. In this case, the applied force $\mathbf{f}$ can be decomposed into a reactive part and an active part,

$$\mathbf{f} = \mathbf{r} + k\mathbf{x}. \tag{4.48}$$

With this analogy in mind, we now see that a similar decomposition applies for the stress of an incompressible material.

Recall that to derive the relationship between the stress and the deformation in hyperelastic solids, we started from the following expression for the change in stored energy

$$dW = \operatorname{tr}(\mathsf{S}\,\mathsf{dF}). \tag{4.49}$$

It turns out that for incompressible solids, the work done dW is the same for the stress S and for the stress $\mathsf{S} + p\,\mathsf{F}^{-1}$, where p is an arbitrary scalar. We can check this statement directly, as follows:

$$\begin{aligned}
\operatorname{tr}[(\mathsf{S} + p\,\mathsf{F}^{-1})\mathsf{dF}] &= \operatorname{tr}(\mathsf{S}\mathsf{dF}) + p\operatorname{tr}(\mathsf{F}^{-1}\mathsf{dF}) \\
&= \operatorname{tr}(\mathsf{S}\,\mathsf{dF}) + p\,d(\det \mathsf{F}) \\
&= \operatorname{tr}(\mathsf{S}\,\mathsf{dF}) + 0, \tag{4.50}
\end{aligned}$$

because $\det \mathsf{F} = 1$. Here we used Jacobi's identity (4.32), which applies for the derivative of the determinant with respect to any scalar field and thus applies for the full differential

$$(\det \mathsf{F}) \operatorname{tr}\left(\mathsf{F}^{-1}\mathsf{dF}\right) = d(\det \mathsf{F}). \tag{4.51}$$

Thus, by enforcing incompressibility, we introduce an arbitrariness in the stress. When we add the term $p\,\mathsf{F}^{-1}$ to S, the work done is the same, just like when we add the term $\mathbf{r}$ to $\mathbf{f}$ in the spring example. Hence, we write

$$\mathsf{S} = -p\,\mathsf{F}^{-1} + \mathsf{S}_a(\mathsf{F}), \qquad \boldsymbol{\sigma} = -p\,\mathsf{I} + \boldsymbol{\sigma}_a(\mathsf{B}), \tag{4.52}$$

where we used $\boldsymbol{\sigma} = J^{-1}\mathsf{F}\mathsf{S}$, with $J = 1$. In both expressions, the p term is the reactive part of the stress. The last term, in analogy to $k\mathbf{x}$, is the active stress, determined by the deformation $\boldsymbol{\sigma}_a = \beta_0 \mathsf{I} + \beta_1 \mathsf{B} + \beta_2 \mathsf{B}^2$, where now, from (4.39),

$$\beta_0 = 2\frac{\partial W}{\partial I_3} = 0, \qquad \beta_1 = 2(W_1 + I_1 W_2), \qquad \beta_2 = -2W_2. \tag{4.53}$$

Here $\beta_0 = 0$ because W depends on I_1 and I_2 only for incompressible materials, not on I_3, which is always fixed at $I_3 = 1$. In conclusion, the general stress-strain relationship for incompressible isotropic hyperelastic materials is

$$\boldsymbol{\sigma} = -p\,\mathsf{I} + 2(W_1 + I_1 W_2)\mathsf{B} - 2W_2 \mathsf{B}^2, \tag{4.54}$$

where $p = p(\mathbf{x})$ is an arbitrary scalar (to be determined later). Using the Cayley-Hamilton theorem, it can also be written as

$$\boldsymbol{\sigma} = -\bar{p}\,\mathsf{I} + 2W_1 \mathsf{B} - 2W_2 \mathsf{B}^{-1}, \tag{4.55}$$

where $\bar{p} = p - 2W_2 I_2$ is also arbitrary. Also, when W is seen as a function of the stretches, the principal Cauchy stresses now read

$$\sigma_i = -p + \lambda_i \frac{\partial W}{\partial \lambda_i}, \qquad i = 1, 2, 3, \tag{4.56}$$

which can be checked by similar computation to the compressible case.

4.7 Examples of Stored-Energy Functions

Many different stored-energy functions are available in the literature to model the behaviour of rubber-like solids and other soft materials. Here we present some examples for *incompressible* isotropic elasticity based on the use of the invariants I_1, I_2, and of the stretches $\lambda_1, \lambda_2, \lambda_3$ (subject to the constraint $\lambda_1 \lambda_2 \lambda_3 = 1$).

A basic stored-energy function, known as the *neo-Hookean material*, has the form

$$W = \frac{\mu_0}{2}(I_1 - 3), \tag{4.57}$$

4.7 Examples of Stored-Energy Functions

where μ_0 (>0) is a material constant referred to as the *shear modulus* of the material in the reference configuration. This is a prototype model for rubber elasticity. According to (4.55), the associated Cauchy stress is

$$\boldsymbol{\sigma} = -p\mathbf{I} + \mu_0 \mathbf{B}. \tag{4.58}$$

With this stress-strain relationship, we can predict how a neo-Hookean material should react to applied forces. Then by collecting experimental data, we can confront predictions to experiments and evaluate the goodness of the model. For this model, we can only adjust the material parameter μ_0 to get a closer fit between experimental and theoretical curves. The next models have two material parameters and might allow for better fits.

One such model is the *Mooney-Rivlin material*, with stored-energy function

$$W = \tfrac{1}{2}C_1(I_1 - 3) + \tfrac{1}{2}C_2(I_2 - 3), \tag{4.59}$$

where C_1 (>0) and C_2 (≥ 0) are constants. We note that this class of models includes the neo-Hookean material, with the specialisation $C_1 = \mu_0$, $C_2 = 0$. The corresponding Cauchy stress is calculated from (4.55) as

$$\boldsymbol{\sigma} = -p\mathbf{I} + C_1 \mathbf{B} - C_2 \mathbf{B}^{-1}. \tag{4.60}$$

Another two-parameter model is the *Gent material*. It is often used to describe solids which stiffen rapidly as they are stretched:

$$W = -\frac{\mu_0 J_m}{2} \ln\left(1 - \frac{I_1 - 3}{J_m}\right), \tag{4.61}$$

where μ_0 and J_m are positive constants. For this material, the Cauchy stress is

$$\boldsymbol{\sigma} = -p\mathbf{I} + \frac{\mu_0 J_m}{J_m + 3 - I_1} \mathbf{B}. \tag{4.62}$$

It shows that the stress blows up when the stretches increase to the point where $I_1 = \lambda_1^2 + \lambda_2^2 + \lambda_3^2$ comes near to $J_m + 3$. Hence, J_m is a "stiffening" parameter: the smaller it is, the earlier the strain-stiffening effect occurs.

Finally, the *Fung material* is a good candidate to model the behaviour of biological soft (isotropic) tissues:

$$W = \frac{\mu_0}{2b}\left[e^{b(I_1-3)} - 1\right], \tag{4.63}$$

where μ_0 and b are positive constants. The Cauchy stress is

$$\boldsymbol{\sigma} = -p\mathbf{I} + \mu_0 e^{b(I_1-3)} \mathbf{B}. \tag{4.64}$$

Here b is also a "stiffening" parameter: the larger it is, the larger the stress is for a given strain. In contrast with the Gent model, there is no upper bound for the stretches.

An example of a stored-energy function for incompressible materials expressed as a function of the stretches is that of the *Ogden material*, given by

$$W = \sum_{n=1}^{N} \frac{\mu_n}{\alpha_n} \left(\lambda_1^{\alpha_n} + \lambda_2^{\alpha_n} + \lambda_3^{\alpha_n} - 3 \right), \tag{4.65}$$

where N is a positive integer and μ_n and α_n are material constants such that

$$\mu_n \alpha_n > 0, \qquad n = 1, 2, \ldots, N. \tag{4.66}$$

From (4.56), the principal Cauchy stresses are calculated as

$$\sigma_i = -p + \sum_{n=1}^{N} \mu_n \lambda_i^{\alpha_n}, \qquad i \in \{1, 2, 3\}. \tag{4.67}$$

This model is most popular for curve-fitting exercises because its accuracy increases rapidly with N. For instance, when choosing to fit an experimental stress-stretch curve with a three-term Ogden material, we have six parameters at our disposal (α_1, α_2, α_3, μ_1, μ_2, μ_3) to fit the theoretical curve to the experimental curve. However, here the curve-fitting exercise is a nonlinear optimisation problem, and it might give several sets of best-fit parameters (non-uniqueness). In that case, we must give up the model because we do not know which set to choose.

Chapter 5
Some Solved Problems of Nonlinear Elasticity

In this final chapter, we look at some problems that we can solve explicitly in nonlinear elasticity. We begin by exploring the difference between homogeneous and inhomogeneous deformations, which are crucial concepts for understanding material responses under different conditions. Homogeneous deformations, such as uni-axial tension or simple shear, maintain uniform strain distributions throughout a material and satisfy the equations of equilibrium automatically. They provide vital formulas for testing soft solids. Inhomogeneous deformations, such as torsion or inflation, provide other testing scenarios essential for determining the strain energy functions of incompressible materials.

Additionally, we explore the fascinating world of electroactive membranes, where elasticity couples with electrostatics to develop advanced applications like soft sensors and artificial muscles. We conclude with an examination of the Biot instability, an essential phenomenon modelling how elastic bodies, when compressed sufficiently, develop wrinkles on their surface.

5.1 Homogeneous and Inhomogeneous Deformations

For homogeneous deformations (defined in Sect. 2.9), the equations of equilibrium in the absence of body forces are automatically satisfied. That is because they read $\text{Div}\,\mathsf{S} = \mathbf{0}$ in general, but here F, F^T, F^{-1} are constant, and thus I_1, I_2, I_3 and W_1, W_2, W_3 are constant as well, so that the constitutive relation leads to a constant S, making the equations of equilibrium a trivial identity.

Why are homogeneous deformations so important to the testing of soft solids? The reason lies in one of the major theorems of nonlinear elasticity, which states that

> A deformation that can be maintained in every compressible, isotropic, hyperelastic material by the application of surface tractions alone is necessarily a *homogeneous* deformation.

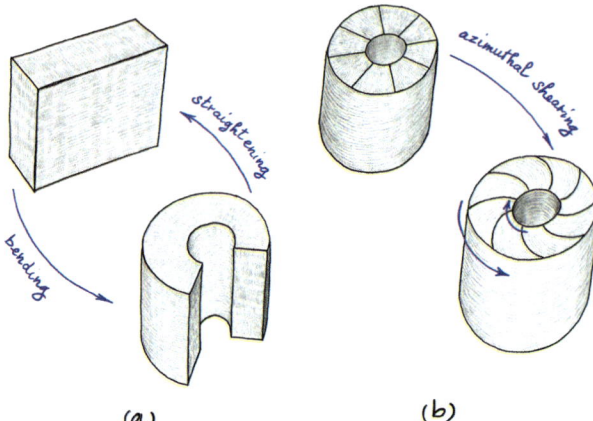

Fig. 5.1 Three inhomogeneous deformations that are universal to all isotropic, incompressible, hyperelastic solids. (**a**) A rectangular block can be bent into a circular cylindrical sector by applying forces and moments on its end faces. Conversely, a circular cylindrical sector can be straightened by a forces and moments. (**b**) In azimuthal shearing, the inner face of a tube is twisted, while the outer face remains fixed (or is twisted by a different amount)

We do not present a proof of this theorem in this introductory course, but the interested reader can find it in more advanced textbooks. This theorem explains why homogeneous deformations are crucial to proper testing protocols: if we were to choose another type of deformation, then we would restrict the possible choices for W, because not all W would be able to model that inhomogeneous deformation.

However, there exist *inhomogeneous* deformations that can be maintained in every *incompressible*, isotropic, hyperelastic material by the application of surface tractions alone. These include the bending of a rectangular block into a circular sector (and its inverse, the straightening of a circular sector into a rectangular block), the torsion of a cylinder or tube, the inflation and eversion of a spherical or cylindrical shell, and the azimuthal shearing of a tube; see Fig. 5.1 for some illustrations.

There are thus more tests that can be implemented to determine the strain energy function of a given incompressible solid. We begin this chapter by modelling the torsion of a solid cylinder and the inflation of a spherical shell.

5.2 Simple Tension Testing

The *simple tension* test is the most widely available method of testing for soft materials. A cuboid sample is clamped at both ends, and a tensile machine imposes a displacement of the clamps; see Fig. 5.2, while measuring the force F and the stretch λ during the extension. By dividing the force with the original cross-sectional area A

5.2 Simple Tension Testing

Fig. 5.2 The deformation of a cuboid under uni-axial tension created by the application of a force of magnitude F. The cross-sectional changes from A in $\mathcal{B}_r$ to a in $\mathcal{B}_c$

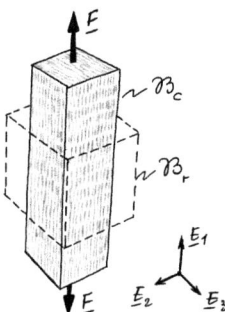

of the sample, we gain access to the nominal stress S_{11}. This test has the advantage that quite large stretches can be achieved for highly deformable soft solids such as rubber and silicone.

Call λ the stretch in the direction of tension: $\lambda = \lambda_1$, say. By symmetry and isotropy, the lateral stretches are equal: $\lambda_2 = \lambda_3$. By the incompressibility condition, $\lambda_1 \lambda_2 \lambda_3 = \lambda \lambda_2^2 = 1$. It follows that the principal stretches of simple extension are

$$\lambda_1 = \lambda, \qquad \lambda_2 = \lambda^{-1/2}, \qquad \lambda_3 = \lambda^{-1/2}. \tag{5.1}$$

In turn, we find

$$[F] = \begin{bmatrix} \lambda & 0 & 0 \\ 0 & \lambda^{-1/2} & 0 \\ 0 & 0 & \lambda^{-1/2} \end{bmatrix}, \quad [B] = \begin{bmatrix} \lambda^2 & 0 & 0 \\ 0 & \lambda^{-1} & 0 \\ 0 & 0 & \lambda^{-1} \end{bmatrix}, \quad [B^{-1}] = \begin{bmatrix} \lambda^{-2} & 0 & 0 \\ 0 & \lambda & 0 \\ 0 & 0 & \lambda \end{bmatrix}, \tag{5.2}$$

so that the invariants I_1, I_2, I_3 are

$$I_1 = \lambda^2 + 2\lambda^{-1}, \qquad I_2 = \lambda^{-2} + 2\lambda, \qquad I_3 = 1. \tag{5.3}$$

In simple tension, we apply a tensile force and no lateral forces: $\sigma_{11} \neq 0$ and $\sigma_{22} = \sigma_{33} = 0$, so that by (4.55),

$$\sigma_{11} = -p + 2W_1 \lambda^2 - 2W_2 \lambda^{-2},$$
$$\sigma_{22} = \sigma_{33} = -p + 2W_1 \lambda^{-1} - 2W_2 \lambda = 0. \tag{5.4}$$

Subtracting one equation from the other, we find that in *simple tension*

$$\sigma_{11} = 2(\lambda^2 - \lambda^{-1})W_1 - 2(\lambda^{-2} - \lambda)W_2 = 2(\lambda^2 - \lambda^{-1})(W_1 - \lambda^{-1} W_2). \tag{5.5}$$

Here, λ is constant, and thus I_1, I_2 and W_1, W_2 are also constants, and so is σ. It follows that the equation of equilibrium, $\operatorname{div} \sigma = \mathbf{0}$, is a trivial identity.

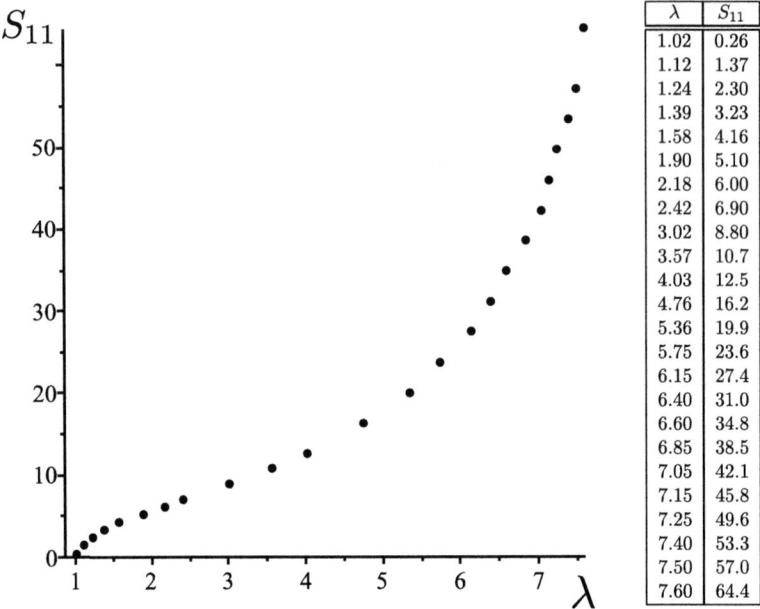

Fig. 5.3 The foundational 1947 data from Treloar for the nominal stress component S_{11} (in N/mm^2) vs the stretch λ in the tensile test of vulcanised rubber

We may also compute the nominal tensile stress S_{11} from the connection $\mathbf{S} = \mathbf{F}^{-1}\boldsymbol{\sigma}$, as $S_{11} = \lambda^{-1}\sigma_{11}$, or

$$S_{11} = 2(\lambda - \lambda^{-2})(W_1 - \lambda^{-1}W_2). \tag{5.6}$$

Figure 5.3 shows representative simple tension data for a strip of vulcanised natural rubber, as collected by Treloar in 1947. It shows the nominal tensile stress S_{11} plotted on the vertical axis against the stretch λ. As noted in Chap. 1, the two changes of curvature are characteristic of the tensile response for rubber. In fact, the vertical data of our tabletop experiment can easily be converted to S_{11}, simply by considering that each additional load was due to the addition of 0.5 l of water (hence a force of $mg = 4.9$ N), so that each additional nominal stress was of the order of 0.245 N/mm^2 (the rubber band used was originally 20 mm wide and 1 mm thick).

Let us have a closer look at the predictions of each model in turn and see how they compare to the data. First we find that the *neo-Hookean model* (4.57) gives the following tensile stress-stretch relationships

$$\sigma_{11} = \mu_0(\lambda^2 - \lambda^{-1}), \qquad S_{11} = \mu_0(\lambda - \lambda^{-2}). \tag{5.7}$$

However, when we plot the function $\lambda - \lambda^{-2}$ for $\lambda \in [1, 8]$, we find that it does not change curvature. This can also be seen from the derivative $dS_{11}/d\lambda = \mu_0(1 +$

5.2 Simple Tension Testing

$2\lambda^{-3}$), which is always positive (monotone increasing curve) but decreases with λ (smaller and smaller slope). Adjusting the parameter μ_0 will merely change the slope at the origin, and we conclude that the neo-Hookean model gives a poor fit and cannot capture the full mechanical behaviour of the rubber sample.

For the *Mooney-Rivlin model* (4.59), we find the stress-stretch connections

$$\sigma_{11} = (\lambda^2 - \lambda^{-1})(C_1 + C_2\lambda^{-1}), \qquad S_{11} = (\lambda - \lambda^{-2})(C_1 + C_2\lambda^{-1}), \qquad (5.8)$$

which might be more versatile for the curve-fitting to the data, having one more adjustable constant compared to the neo-Hookean model. One good way to test whether the material's tensile response is accurately depicted by the Mooney-Rivlin model is to re-scale the data and plot the quantity $y = S_{11}/(\lambda - \lambda^{-2})$ on the vertical axis and $x = \lambda^{-1}$ on the horizontal axis. In that way, the Mooney-Rivlin model predicts that according to $(5.8)_2$, y should be related to x through

$$y = C_1 + C_2 x, \qquad (5.9)$$

and therefore we should get a straight line in the (x, y)-plane. Then C_1 would be the intercept of the line and C_2, its slope. This re-scaling produces what is called the *Mooney plot*.

Figure 5.4a shows the Mooney plot for the Treloar data. Going from $x = 1$ (no stretch) towards $x = 0$ (increasing stretch $\lambda = x^{-1}$), we see that the scaled data is linear only in the beginning of the tensile test, when λ^{-1} goes from 1 to 0.46, which

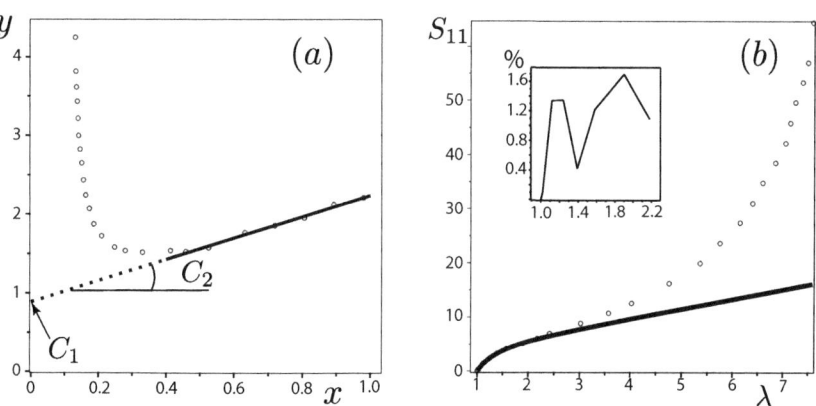

Fig. 5.4 (a) Mooney plot of the Treloar data for rubber: quantity $y = S_{11}/(\lambda - \lambda^{-2})$ against $x = \lambda^{-1}$, revealing linear correspondence between x and y for the first seven or eight data points. From that line, we obtain the C_1 (intercept) and C_2 (slope) parameters of the Mooney-Rivlin model. (b) Corresponding prediction of the Mooney-Rivlin material in simple tension, superposed on the Treloar data for rubber (tensile nominal stress S_{11} in N/mm^2). The inset shows that the error is small when the fitting is done for the first eight points only (up to $\lambda = 2.2$, that is, an extension of 120%)

corresponds to the stretch range $1 \le \lambda \le 2.2$. In that range, the Mooney-Rivlin model does a good job of predicting the tensile response, with relative errors below 1.7%; see inset of Fig. 5.4b, which also shows the corresponding Mooney-Rivlin prediction superposed on the data. Hence, if we are only interested in predicting the behaviour of rubber in the small-to-moderate range, then the Mooney-Rivlin model is a good candidate.

Similar to the neo-Hookean model, the Mooney-Rivlin model does not change curvature in the tensile response. However, the Mooney plot shows that the neo-Hookean model cannot capture the behaviour of rubber in any part of the tensile response, because it provides only a horizontal line at $y = C_1 = \mu_0$, which cannot be made to fit any region of the Mooney plot, in contrast to the Mooney-Rivlin model. Finally, we see that the Mooney plot displays an "upturn" when we go beyond the small-to-moderate range ($x < 0.46$), which calls for a strain-stiffening model.

The *Gent model* provides a stiffening response, as we can see from the expression for the tensile stress, which according to (4.62), (5.3), and (5.5) reads

$$\sigma_{11} = \mu_0 \frac{\lambda^2 - \lambda^{-1}}{1 - \frac{\lambda^2 + 2\lambda^{-1} - 3}{J_m}}, \qquad S_{11} = \mu_0 \frac{\lambda - \lambda^{-2}}{1 - \frac{\lambda^2 + 2\lambda^{-1} - 3}{J_m}}. \qquad (5.10)$$

These expressions clearly show that there is a vertical asymptote in the response graph, located at a maximum axial stretch $\lambda = \lambda_m$ defined by $\lambda_m^2 + 2\lambda_m^{-1} - 3 = J_m$. By using this model for the curve fitting exercise, we find that the predictive power of the Gent model is excellent in the large extension regime where the stiffening occurs, as expected, as shown in Fig. 5.5a. However, it is not so good in the small-to-moderate regime, where the error can be as high as 40%.

This observation prompts us to propose the following combination of the Mooney-Rivlin model, which performs well in the small-to-moderate regime, and the Gent model, which performs well in the stiffening regime, as

$$W = -\frac{C_1 J_m}{2} \ln\left(1 - \frac{I_1 - 3}{J_m}\right) + \frac{C_2}{2}(I_2 - 3), \qquad (5.11)$$

with three adjustable parameters C_1, J_m, C_2. For this model, the predicted tensile stress is

$$\sigma_{11} = C_1 \frac{\lambda^2 - \lambda^{-1}}{1 - \frac{\lambda^2 + 2\lambda^{-1} - 3}{J_m}} + C_2(\lambda - \lambda^{-2}),$$

$$S_{11} = C_1 \frac{\lambda - \lambda^{-2}}{1 - \frac{\lambda^2 + 2\lambda^{-1} - 3}{J_m}} + C_2(1 - \lambda^{-3}). \qquad (5.12)$$

5.2 Simple Tension Testing

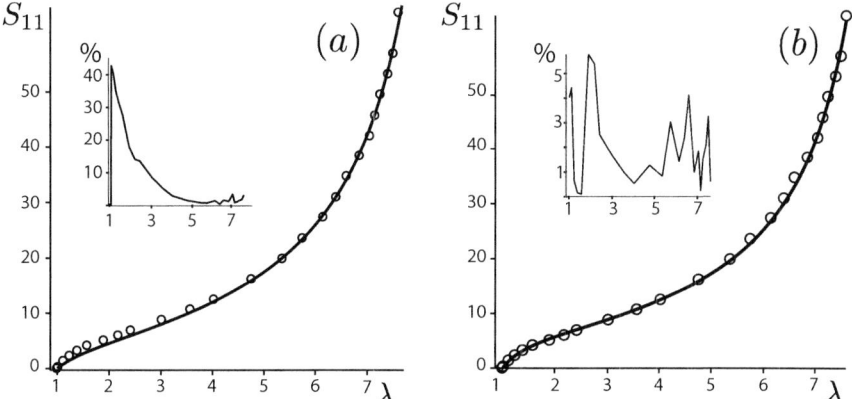

Fig. 5.5 Simple tension data (circles) for rubber fitted to (**a**) the Gent model, (**b**) the Gent+Mooney-Rivlin model (continuous plots). The nominal stress is in N/mm^2. The insets show that the relative error is quite large in the small-to-moderate range for the Gent model (up to 40%), while it is consistently low (less than 5.5%) in the entire range for the latter model

The fit to the data is then excellent over the whole range of recorded stretches; see Fig. 5.5b, showing that the relative error is less than 5.5% over the entire range of deformation.

As we can see, modelling through curve-fitting is an art as well as a science. Here, we just touched upon that subject, and there are many aspects that we did not develop. We now mention a couple.

We relied on built-in procedures for the parameter optimisation, which can be found in all available scientific software codes: for example, in Maple and Mathematica, we may use the commands `NonlinearFit` and `NonlinearModelFit`, respectively; in Matlab, we may use the Curve Fitting Tool graphic user interface, etc. By default, the optimisation minimises the sum of the absolute errors squared, $\sum |f(\lambda_i) - y_i|^2$, where y_i is the stress measured for each λ_i, and f is the stress function to be fitted. For our last example, f is taken from (5.12) and the procedure adjusts C_1, C_2 and J_m to minimise that sum. This approach is problematic, because the optimisation procedure will return different parameters depending on whether the function is σ_{11} or S_{11}, even though the material parameters in W are independent of the subsequent choice of stress measure. To avoid this problem, we may minimise the sum of the relative errors squared, $\sum |f(\lambda_i)/y_i - 1|^2$, which is independent of the choice of stress measure. This is easily implemented, by using the `weighting` option.

Another aspect that we did not consider is that so far, our models have been *descriptive* only. A good model must not only fit reasonably well the data from one test, it must also be *predictive*. Ideally, we should perform several types of tests on a given material: for example, a uni-axial tensile test, for which we select a model giving a good curve fitting, and an equi-biaxial test, or simple shear test, torsion test, etc., for which we may confront the data to the predicted curve.

5.3 Simple Shear Testing

For simple tension, the directions of the principal axes of stretch do not change as the magnitude of the stretch is varied. We now consider the predictions of the theory with a deformation for which the orientation of the principal axes of stretch *does* change: the *simple shear deformation*,

$$x_1 = X_1 + KX_2, \qquad x_2 = X_2, \qquad x_3 = X_3. \tag{5.13}$$

Simple shear is an important deformation because it arises locally in many problems of practical and theoretical interest. According to the standard testing protocols, this deformation is achieved by cutting a rectangular sample of material with edges parallel to the X_i axes in the reference configuration, gluing its opposite faces normal to the X_2 axis to rigid platens, and moving one platen parallel to the other in the 1 direction (see Fig. 5.6) while recording the displacement d and the required shear force F_{12}. Then the amount of shear is $K = d/h$ where h is the current height of the sample, and the nominal shear stress component is $S_{12} = F_{12}/A$, where A is the area of the glued surface in the reference configuration.

From the simple shear deformation (5.13), we find

$$[\mathbf{F}] = \begin{bmatrix} 1 & K & 0 \\ 0 & 1 & 0 \\ 0 & 0 & 1 \end{bmatrix}, \qquad [\mathbf{F}^{-1}] = \begin{bmatrix} 1 & -K & 0 \\ 0 & 1 & 0 \\ 0 & 0 & 1 \end{bmatrix}. \tag{5.14}$$

Hence, $J = \det \mathbf{F} = 1$, emphasising that simple shear is an *isochoric* deformation, always compatible with the constraint of incompressibility. Also, the height of the sample is unchanged in simple shear, because from (5.13)$_2$, $h = H$. Then the amount of shear $K = d/h = d/H$ is easy to compute from the measurement of the displacement.

Fig. 5.6 Simple shear and tractions on the faces with normals $\mathbf{e}_2$ and $\mathbf{e}_3$

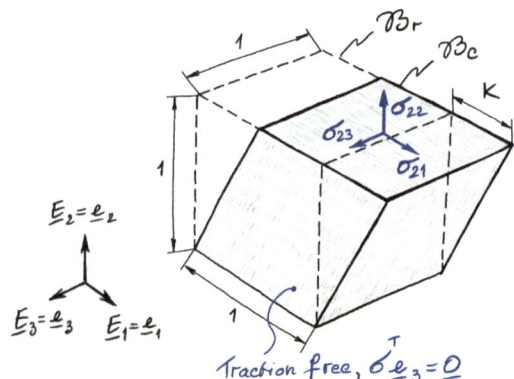

5.3 Simple Shear Testing

Note also that, by Nanson's formula,

$$\mathbf{e}_2 da = \mathbf{F}^{-T}\mathbf{E}_2 dA = \begin{bmatrix} 1 & 0 & 0 \\ -K & 1 & 0 \\ 0 & 0 & 1 \end{bmatrix} \begin{bmatrix} 0 \\ 1 \\ 0 \end{bmatrix} dA = \begin{bmatrix} 0 \\ 1 \\ 0 \end{bmatrix} dA = \mathbf{E}_2 dA, \quad (5.15)$$

showing that the areas of surface elements in the horizontal planes remain unchanged. Hence, the area of the glued surfaces is unchanged ($A = a$, the current area), so that the Cauchy shear stress component, $\sigma_{12} = F_{12}/a = F_{12}/A$, is easy to evaluate from F_{12}. Notice that $\sigma_{12} = S_{12}$: the Cauchy and the nominal shear stress components are equal.

From the simple shear deformation (5.13), we find

$$[\mathbf{B}] = \begin{bmatrix} 1+K^2 & K & 0 \\ K & 1 & 0 \\ 0 & 0 & 1 \end{bmatrix}, \quad [\mathbf{B}^{-1}] = \begin{bmatrix} 1 & -K & 0 \\ -K & 1+K^2 & 0 \\ 0 & 0 & 1 \end{bmatrix}, \quad (5.16)$$

so that the invariants I_1, I_2, I_3 are

$$I_1 = I_2 = 3 + K^2, \qquad I_3 = 1. \quad (5.17)$$

Using (4.55), we find the following components for σ,

$$\sigma_{11} = -p + 2W_1(1 + K^2) - 2W_2, \qquad \sigma_{12} = 2(W_1 + W_2)K,$$
$$\sigma_{22} = -p + 2W_1 - 2W_2(1 + K^2), \qquad \sigma_{13} = \sigma_{32} = 0,$$
$$\sigma_{33} = -p + 2(W_1 - W_2). \quad (5.18)$$

First of all, we notice that W depends on K^2 only, according to (5.17), so that W_1 and W_2 are functions of K^2 also. Because $\sigma_{12} = 2(W_1 + W_2)K$, it follows that the *shear stress* σ_{12} is related to the amount of shear K in an *odd* manner: $\sigma_{12}(-K) = -\sigma_{12}(K)$ or, in other words, the shear stress required to produce the amount of shear K is the opposite of the shear stress required to produce the amount of shear $-K$. We emphasise this important result by defining the *nonlinear shear modulus* μ:

$$\sigma_{12} = \mu K, \quad \text{where} \quad \mu = \mu(K^2) = 2\left(\frac{\partial W}{\partial I_1} + \frac{\partial W}{\partial I_2}\right). \quad (5.19)$$

We also define the *initial shear modulus* μ_0, as

$$\mu_0 = \mu(0) = 2\left(\frac{\partial W}{\partial I_1} + \frac{\partial W}{\partial I_2}\right)_{I_1 = I_2 = 3}, \quad (5.20)$$

which is the slope of the tangent to the $\sigma_{12} - K$ curve at the origin.

For the *neo-Hookean* and the *Mooney-Rivlin materials*, $\mu(K^2)$ is a *constant*:

$$\mu(K^2) = \mu_0, \quad \text{and} \quad \mu(K^2) = C_1 + C_2, \tag{5.21}$$

respectively, and the nonlinear and initial shear moduli coincide. For these materials, the $\sigma_{12} - K$ curve is a straight line, with constant slope.

For the *Gent, Fung*, and *Gent+Mooney-Rivlin materials*, we find that $\mu(K^2)$ is

$$\mu_0 \frac{J_m}{J_m - K^2}, \quad \mu_0 \exp(bK^2), \quad C_1 \frac{J_m}{J_m - K^2} + C_2, \tag{5.22}$$

respectively. The corresponding initial shear moduli $\mu(0)$ are μ_0, μ_0, $C_1 + C_2$, respectively. For these three models, the $\sigma_{12} - K$ curve has an ever increasing slope, because μ increases with K.

In Fig. 5.7, shear stress data is reported for porcine brain tissue. Some authors have tried to fit this type of curve to a "power law" of the form $\sigma_{12} = aK^b$, where a and b are fitting parameters. Clearly, this is unphysical, because it does not respect the odd relationship between shear stress and amount of shear (unless b is an odd integer).

Instead, we can look closely at the curve and notice that in the range $0 \le K \le 1$, the stress-shear relationship is almost *linear*. A simple linear curve-fitting exercise gives a good coefficient of determination here, $R^2 = 0.994$.

We can thus conclude that the neo-Hookean and the Mooney-Rivlin materials are good candidates to model the behaviour of this sample, because they both predict a

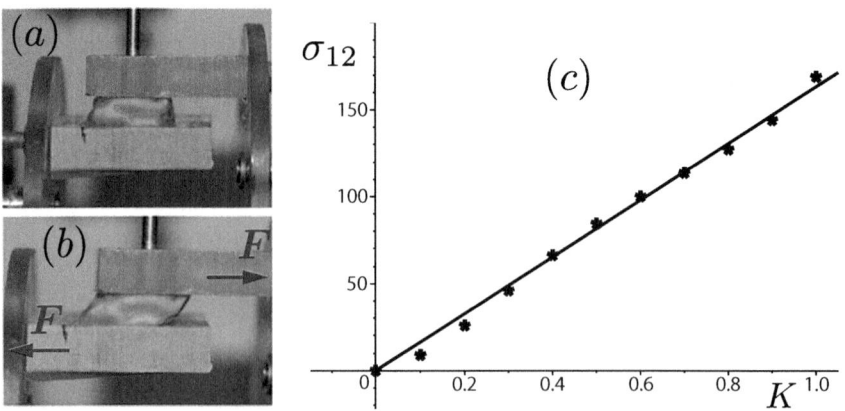

Fig. 5.7 Large simple shear of pig brain matter. (**a**) A sample of height $H = 1$ cm, width $W = 4$ cm, and length $L = 4$ cm is glued on its large faces to rigid platens. (**b**) The platens are then subjected to a shear force F (in N). (**c**) The Cauchy shear stress component $\sigma_{12} = F/A$ (in Pa), where $A = WL$, is plotted against the amount of shear $K = d/H$, where d is the horizontal displacement of the platens. (Reproduced with permission from [1])

5.3 Simple Shear Testing

linear plot. However, we do not have enough information yet to favour one model over the other. By computing the slope of the straight line, we find that the brain matter sample has a shear modulus of about $\mu = \mu_0 = 180$ Pa, putting it in the category of *extremely soft solids*.

Now we go back to the stress components (5.18) and notice the connection

$$\sigma_{11} - \sigma_{22} = K\sigma_{12}. \tag{5.23}$$

This relationship is an example of a *universal relation*, i.e. a connection between the stress components that is the same for all incompressible, isotropic, and hyperelastic solids (independent of the strain energy function).

It follows from the universal relation (5.23) that the normal stresses σ_{11} and σ_{22} can never be equal in simple shear. In particular, they can never be zero simultaneously. This phenomenon is called the *Poynting effect*.

Now we use boundary conditions to fix the value of the Lagrange multiplier p, required to determine all the stress components in (5.18). Specifically, we focus the analysis on the case of *plane stress*, where there is no force applied to the front and back faces: $\sigma_{33} = 0$. According to (5.18), this leads to $p = 2(W_1 - W_2)$ and to the following expressions for the normal stresses (the components of σ in the directions normal to the faces of the sample):

$$\sigma_{11} = 2W_1 K^2, \qquad \sigma_{22} = -2W_2 K^2. \tag{5.24}$$

Again we see that the normal stresses σ_{11} and σ_{22} cannot be zero simultaneously. (Otherwise we would have $W_1 = W_2 = 0$ and W would depend neither on I_1 nor on I_2).

We conclude that in general, *normal stresses* are required in addition to shear stress to maintain the shape of the sheared block. Hence, simple shear is not achieved by a shear stress alone. The necessity for normal forces is an example of the *Kelvin effect*. In other words, it is not enough to just pull on one platen while keeping the other one fixed. The upper platen must also be restricted from moving upwards or downwards, and forces must be applied to counterbalance the normal stresses. Practically, this is achieved by guiding the platen horizontally.

Notice that the Kelvin effect is *nonlinear*: if K was infinitesimal, as in Linear Elasticity, then we would neglect terms in K^2 and higher powers, and from (5.24), σ_{11}, σ_{22} would both be zero.

Going back to our example where we tried to model the behaviour of porcine brain tissue, we saw that σ_{12} is related to K in a linear manner. Hence, W can be of the neo-Hookean type, for which we find that

$$\sigma_{22} = 0, \tag{5.25}$$

or of the Mooney-Rivlin type, for which we find that

$$\sigma_{22} = -2C_2 K^2, \tag{5.26}$$

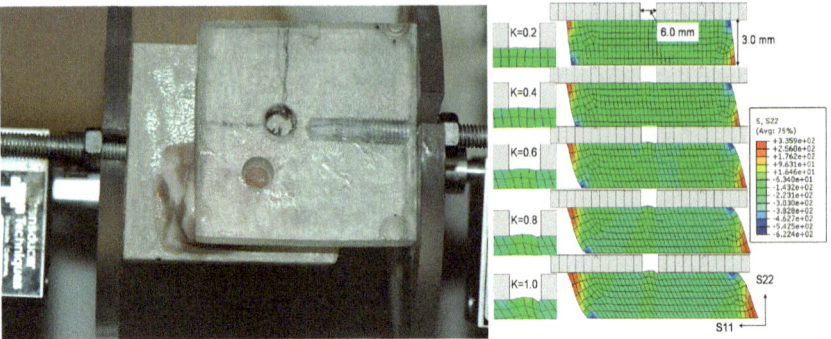

Fig. 5.8 Normal stress arising in the simple shear of porcine brain matter leads it to bulge out of a hole drilled in one of the platens. Left: experiments, Right: computer simulations. (Reproduced with permission from [1])

according to (4.57), (4.59), and (5.19) (provided the plane stress assumption is valid). It follows that no forces normal to the platens appear in the neo-Hookean case, while they do appear in the Mooney-Rivlin case. In the experiments, we drilled a hole in the upper platen and observed that the brain matter bulged through the hole; see Fig. 5.8. It follows that $\sigma_{22} \neq 0$, and thus the neo-Hookean model is ruled out. In fact, the Gent model (4.61) and the Fung model (4.63) must also be discarded to model this material, because they both give $W_2 = 0$ and thus, $\sigma_{22} = 0$ by $(5.24)_2$. In conclusion, the brain matter of the sample must be described by the Mooney-Rivlin model.

Note that to stop the brain matter from expanding along $\mathbf{e}_2$, we must supply a normal force $\sigma_{22} \times a$ directed along $-\mathbf{e}_2$, so that $\sigma_{22} < 0$, which confirms that $C_2 > 0$, as we had assumed when we introduced the Mooney-Rivlin material.

Finally, we compute the *forces acting on the slanted faces* of the sheared sample, if any. We saw in Chap. 2 that the normal to those faces is along $(1, -K, 0)$, so that the unit normal and tangent vectors to the slanted faces are

$$\mathbf{n} = \frac{1}{\sqrt{1+K^2}}\begin{bmatrix} 1 \\ -K \\ 0 \end{bmatrix}, \qquad \mathbf{t} = \frac{1}{\sqrt{1+K^2}}\begin{bmatrix} K \\ 1 \\ 0 \end{bmatrix}, \qquad (5.27)$$

respectively. Then we can compute the normal stress σ and the tangent stress τ on the slanted faces, as $\sigma = \mathbf{n} \cdot \boldsymbol{\sigma}^\mathsf{T}\mathbf{n}$ and $\tau = \mathbf{t} \cdot \boldsymbol{\sigma}^\mathsf{T}\mathbf{n}$. Here we find, using the universal relation (5.23), that (check as an exercise)

$$\sigma = \sigma_{22} - \frac{K}{1+K^2}\sigma_{12}, \qquad \tau = \frac{1}{1+K^2}\sigma_{12}. \qquad (5.28)$$

We see that the forces acting on the slanted faces cannot be zero, so that simple shear cannot be achieved without supplying support to these faces. In other words, only the faces that are perpendicular to $\mathbf{e}_3$ can be free of traction. It follows that the sample must be put inside a "shearing-box" with four walls that will support the other four faces during the simple shear, but this requires the slanted faces of the box to expand as they follow the shear stretch. In effect, this is never implemented in the lab! A sensible alternative would be to use the shear box setup of Chap. 2, where the four walls do follow the four faces of the cube (but this is not simple shear).

If there is no support supplied to the lateral faces, they bend and lead to an inhomogeneous deformation, not described by the simple shear deformation (5.13). To minimise the effect of the inhomogeneities, the standard protocols recommend that the sample should be at least four times wider than it is high, so that the inhomogeneities are localised close to the lateral faces, and do not affect the rest of the deformation too much; see the Finite Element simulations in Fig. 5.8.

5.4 Pure Torsion of an Incompressible Cylinder

The torsion test is quite straightforward to implement in the lab. We first prepare a cylindrical sample of the soft material to be tested and place it in a rheometer machine, by gluing its top and bottom faces to circular plates. The device then applies a torque on the top plate while keeping the bottom one fixed, thus twisting the sample. The machine can be programmed to measure and record the torque and the normal force on the top plate as the twist angle is increased in small increments.

Theoretically, the treatment of the torsion deformation can be done explicitly and analytically and reveals interesting features.

5.4.1 Equilibrium in the Torsion Problem

Here, we check that pure torsion is indeed a solution to the equations of equilibrium for all isotropic, incompressible, and hyperelastic solids. It is best described in a cylindrical coordinate system. The deformation mapping for this case is

$$r = R, \qquad \theta = \Theta + \gamma Z, \qquad z = Z, \tag{5.29}$$

where γ is a constant, the *amount of twist*. Here (R, Θ, Z) and (r, θ, z) are the coordinates in the current and reference configurations, respectively.

Computing the deformation gradient **F** in the cylindrical coordinate system is quite involved. Here, we rely on the following formula, found in textbooks,

$$[\mathbf{F}] = \begin{bmatrix} \dfrac{\partial r}{\partial R} & \dfrac{1}{R}\dfrac{\partial r}{\partial \Theta} & \dfrac{\partial r}{\partial Z} \\ r\dfrac{\partial \theta}{\partial R} & \dfrac{r}{R}\dfrac{\partial \theta}{\partial \Theta} & r\dfrac{\partial \theta}{\partial Z} \\ \dfrac{\partial z}{\partial R} & \dfrac{1}{R}\dfrac{\partial z}{\partial \Theta} & \dfrac{\partial z}{\partial Z} \end{bmatrix}, \tag{5.30}$$

from which it is easy to compute

$$[\mathbf{F}] = \begin{bmatrix} 1 & 0 & 0 \\ 0 & 1 & \gamma R \\ 0 & 0 & 1 \end{bmatrix}, \quad [\mathbf{F}^{-1}] = \begin{bmatrix} 1 & 0 & 0 \\ 0 & 1 & -\gamma R \\ 0 & 0 & 1 \end{bmatrix}. \tag{5.31}$$

Clearly, $J = \det \mathbf{F} = 1$, and thus pure torsion is compatible with incompressibility. Also, we see that one of the components of **F** is γR, not a constant. Hence, torsion is indeed a *non-homogeneous* deformation (Fig. 5.9).

Now we can compute the components of **B** and $\mathbf{B}^{-1}$ and substitute into the stress-deformation relationship $\boldsymbol{\sigma} = -p\mathbf{I} + 2W_1\mathbf{B} - 2W_2\mathbf{B}^{-1}$ to find the non-zero components of the Cauchy stress tensor as

$$\sigma_{rr} = -p + 2W_1 - 2W_2,$$
$$\sigma_{\theta\theta} = -p + 2W_1(1 + \gamma^2 r^2) - 2W_2,$$
$$\sigma_{zz} = -p + 2W_1 - 2W_2(1 + \gamma^2 r^2),$$
$$\sigma_{\theta z} = 2(W_1 + W_2)\gamma r, \tag{5.32}$$

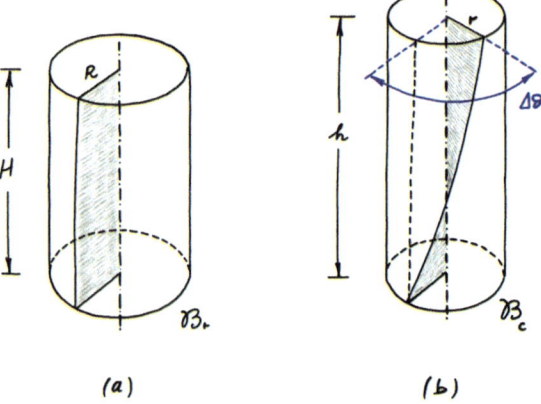

Fig. 5.9 Combined torsion and axial stretching for an incompressible cylinder in its reference (**a**) and current (**b**) configurations

5.4 Pure Torsion of an Incompressible Cylinder

where we are replacing R with r because the Cauchy stress is defined in the current configuration $\mathcal{B}_c$ (this substitution is only formal in this case, because $r = R$). In the process, we also find that

$$I_1 = I_2 = 3 + \gamma^2 r^2,$$

showing that W_1 and W_2 depend on r only.

In general, the equations of equilibrium in the absence of body forces, div $\sigma = \mathbf{0}$, are hard to write down in a non-rectangular coordinate system. From textbooks, we find that in the cylindrical coordinate system (r, θ, z), they read as follows:

$$\frac{\partial \sigma_{rr}}{\partial r} + \frac{1}{r}\frac{\partial \sigma_{r\theta}}{\partial \theta} + \frac{\sigma_{rr} - \sigma_{\theta\theta}}{r} + \frac{\partial \sigma_{rz}}{\partial z} = 0,$$

$$\frac{\partial \sigma_{\theta r}}{\partial r} + \frac{1}{r}\frac{\partial \sigma_{\theta\theta}}{\partial \theta} + \frac{2\sigma_{\theta r}}{r} + \frac{\partial \sigma_{\theta z}}{\partial z} = 0,$$

$$\frac{\partial \sigma_{zr}}{\partial r} + \frac{1}{r}\frac{\partial \sigma_{z\theta}}{\partial \theta} + \frac{\sigma_{zr}}{r} + \frac{\partial \sigma_{zz}}{\partial z} = 0. \tag{5.33}$$

Here they simplify to

$$-\frac{\partial p}{\partial r} + 2\frac{d}{dr}(W_1 - W_2) - 2\gamma^2 r W_1 = 0,$$

$$-\frac{\partial p}{\partial \theta} = 0,$$

$$-\frac{\partial p}{\partial z} = 0. \tag{5.34}$$

The last two equations tell us that $p = p(r)$ only. The first one then becomes an equation giving $p'(r)$. By integrating it, we determine the Lagrange multiplier p such that the equations of equilibrium are satisfied. As we see below, we can formally perform that integration for any W and then satisfy the boundary conditions. Thus, pure torsion is indeed a universal solution for all incompressible isotropic materials.

We find that

$$p(r) = 2[W_1(r) - W_2(r)] - 2\gamma^2 \int^r x W_1(x)dx + C, \tag{5.35}$$

where C is a constant of integration. Then the normal stress component σ_{rr} follows as

$$\sigma_{rr}(r) = 2\gamma^2 \int^r x W_1(x)dx + C. \tag{5.36}$$

To fix C, we consider the case of a full cylinder of radius $A = a$ (the radius remains unchanged by the pure torsion), and we assume that its curved face is free of traction so that $\sigma_{rr}(a) = 0$. This gives

$$\sigma_{rr}(r) = 2\gamma^2 \int_a^r x W_1(x) dx. \tag{5.37}$$

From this expression, we can also compute σ_{zz} by using the expression of $\sigma_{zz} - \sigma_{rr}$ coming from (5.32). We find

$$\sigma_{zz}(r) = 2\gamma^2 \int_a^r x W_1(x) dx - 2\gamma^2 r^2 W_2(r). \tag{5.38}$$

We can use these expressions to test an incompressible material in torsion, and perhaps access its strain-energy function W. To effect pure torsion, we need to keep the top and bottom faces of the cylinder at the same distance one from another and apply a twisting moment (torque). This quantity is

$$M = \int_0^a \int_0^{2\pi} r\sigma_{z\theta} r d\theta dr = 2\pi \int_0^a \sigma_{z\theta} r^2 dr = 4\pi\gamma \int_0^a r^3 (W_1 + W_2) dr, \tag{5.39}$$

because $r\sigma_{z\theta}$ is the moment of the force per unit area, and $rd\theta dr$ is the area of an elementary surface on the top and bottom faces; see Fig. 5.10.

Say, for instance, that we wish to check whether the solid is described by the Mooney-Rivlin strain energy density $W = \tfrac{1}{2}C_1(I_1 - 3) + \tfrac{1}{2}C_2(I_2 - 3)$, for which $2(W_1 + W_2) = C_1 + C_2$. Then we must check if, experimentally, the moment M varies linearly with the amount of twist γ, because in that case, the moment is, according to (5.39)

$$M_{\text{MR}} = \tfrac{1}{2}(C_1 + C_2)\pi a^4 \gamma. \tag{5.40}$$

If the $M - \gamma$ data does not display a linear dependence, then we must turn to other models for an adequate prediction. For example, the Fung model gives $W_1 =$

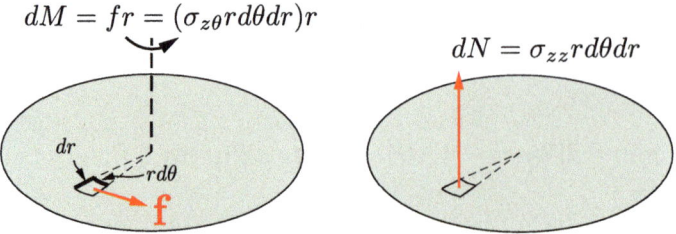

Fig. 5.10 To compute the total moment M and total normal force N applied to the end faces of the cylinder, we decompose each face into the sum of elementary surfaces of area $rd\theta dr$

5.4 Pure Torsion of an Incompressible Cylinder

$\frac{\mu}{2}e^{b(I_1-3)} = \frac{\mu}{2}e^{b\gamma^2 r^2}$, $W_2 = 0$, so that

$$M_{\text{Fung}} = \frac{\mu\pi}{b^2\gamma^3}\left[1 + (ba^2\gamma^2 - 1)e^{b\gamma^2 a^2}\right], \tag{5.41}$$

and we may use that relation to predict a stiffening response in torsion.

Finally, another question arises: Do we need to apply a normal force on the end faces of the cylinder to maintain the pure torsion? To answer it, we compute the normal force on these faces:

$$N = \int_0^a \int_0^{2\pi} \sigma_{zz} r\, d\theta\, dr = 2\pi \int_0^a \sigma_{zz} r\, dr$$

$$= 4\pi\gamma^2 \int_0^a r \left(\int_a^r xW_1(x)dx\right) dr - 4\pi\gamma^2 \int_0^a r^3 W_2(r)dr$$

$$= 4\pi\gamma^2 \left[\frac{r^2}{2}\left(\int_a^r xW_1(x)dx\right)\right]_0^a - 4\pi\gamma^2 \int_0^a \frac{r^2}{2} rW_1(r)dr$$

$$- 4\pi\gamma^2 \int_0^a r^3 W_2(r)dr$$

$$= -2\pi\gamma^2 \int_0^a r^3(W_1 + 2W_2)dr, \tag{5.42}$$

where we used integration by parts. For instance, the Mooney-Rivlin model, for which $W_1 = C_1/2$, $2W_2 = C_2$, predicts that the normal force should be proportional to the squared amount of twist:

$$N_{\text{MR}} = -\tfrac{1}{4}(C_1 + 2C_2)\pi a^4 \gamma^2. \tag{5.43}$$

If the two relationships (5.40) and (5.43) are satisfied by the data, then the cylinder is best described by the Moooney-Rivlin model.

For example, the results of torsion experiments on two samples of gel in Fig. 5.11 display these linear dependencies. The slopes of the two straight lines give access to the material constants C_1 and C_2. Take the slopes of the data for gel A: for the $M - \gamma$ and $N - \gamma^2$ data, we find (see dotted lines) that they are

$$\tfrac{1}{2}(C_1 + C_2)\pi a^4 = \tfrac{0.0075 - 0.00175}{100 - 20}, \quad -\tfrac{1}{4}(C_1 + 2C_2)\pi a^4 = \tfrac{-0.675 - (-0.125)}{12000 - 2000}. \tag{5.44}$$

The radius of the sample is $a = 4$ mm, and thus by solving these simultaneous equations, we arrive at $C_1 = 84$ kPa, $C_2 = 95$ kPa, and the material is completely characterised. In particular, we see that its shear modulus is $\mu_0 = C_1 + C_2 \simeq 180$ kPa, showing that this gel is 1000 times stiffer than the brain matter sample tested in simple shear in Sect. 5.3.

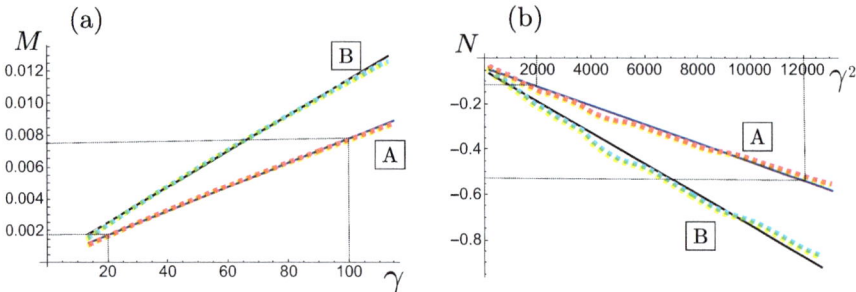

Fig. 5.11 Torsion experiments on two cylindrical samples of gel (gel A: orange data, gel B: green data), both with radius $a = 4$ mm. (**a**) Torque against twist. (**b**) Normal force against squared twist. Linear curve fittings indicate that these two gels behave as Mooney-Rivlin solids

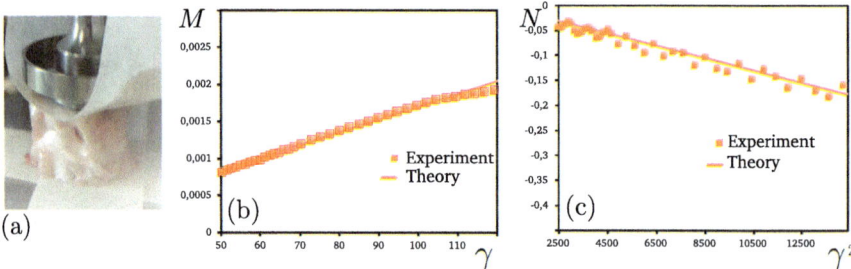

Fig. 5.12 Using a rheometer to characterise brain matter: results of experiments and theoretical predictions. (**a**) A cylindrical sample is harvested from a porcine brain and glued to the platens (with masking tape in between). (**b**) Torque against twist. (**c**) Normal force against squared twist. Here too, linear curve fittings indicate that brain matter behaves as a Mooney-Rivlin solid

When the response is characteristic of the Mooney-Rivlin model, we need two datasets to obtain the two constants C_1 and C_2 from torsion experiments, like in simple shear. In the lab, it is easy to measure σ_{12} in simple shear, but difficult to measure σ_{22}; however, modern rheometers now provide readings not only for the moment M but also for the normal force N in torsion. For brain matter, we can readily conduct torsion experiments, although the protocol is delicate to implement, as brain is very soft, fragile, and slippery. Figure 5.12 shows the data obtained for the torsion of a sample of porcine brain. There we find that $C_1 = 630$ Pa, $C_2 = 470$ Pa, giving a shear modulus $\mu_0 = 1100$ Pa, about six times stiffer than the brain sample tested in simple shear in Sect. 5.3, for which $\mu_0 = 180$ Pa.

Another important conclusion is that a compressive normal force $N_{\text{MR}} < 0$ is necessary to maintain the torsion. If there was no normal force, the cylinder would *lengthen in torsion*, a somewhat counterintuitive phenomenon first discovered by Poynting in 1909 for steel and rubber cords. This is the original "Poynting effect", a denomination that was later extended to describe the expansion of a rectangular sample vertically when subjected to a simple shear horizontally, as we saw in the previous section.

5.4.2 Energy Method for the Poynting Effect in Torsion

The tendency of an isotropic and homogeneous cylinder to elongate during torsion, the so-called Poynting effect, can also be proved by an energy approach. Here the focus is simply to prove that the cylinder elongates, and we do not need to compute the stress, which simplifies significantly the resulting calculations.

The main idea relies upon the *principle of energy minimization*, which states that when an elastic body is subjected to the action of forces, it deforms so as to minimise a suitable "energy". This principle can be easily illustrated with the simple example of Hooke's Law for a spring.

Take a one-dimensional spring, whose state is completely characterised by its length. Call x_0 its stress-free length, and x its length upon the application of a force F, then

$$F = k(x - x_0), \tag{5.45}$$

where the positive constant k is the spring stiffness. The relation above bears the name of "Hooke's law".[1] At equilibrium, from (5.45), we find that the spring has length $x^\star = x_0 + F/k$. This length minimises the *total potential energy* of the spring, which can be defined as follows: First, introduce the spring's *elastic energy*,

$$U = \tfrac{1}{2}k(x - x_0)^2, \tag{5.46}$$

and the mechanical work done by the force F,

$$W_F = F(x - x_0). \tag{5.47}$$

Then the *total potential energy* is defined as

$$E = U - W_F. \tag{5.48}$$

If the force F is prescribed, we can now show that the equilibrium length $x^\star = x_0 + F/k$ is a minimizer of the energy $E(x)$, in the sense that $E(x^\star) \leq E(x)$ for all $x \neq x^\star$. Indeed, we find the minimum of E from the conditions

$$\frac{dE}{dx} = k(x - x_0) - F = 0, \qquad \frac{d^2E}{dx^2} = k \geq 0, \tag{5.49}$$

confirming that $x^\star$ is a minimiser of E, and that the stiffness k is positive.

---

[1] When Hooke established it in the seventeenth century, he safeguarded its discovery by publishing it under the anagram *ceiiinosssttuv* of the Latin sentence *ut tensio, sic vis*, which translates as *as the extension, so the force*.

The same principle can now be used to prove the existence of the Poynting effect in torsion. To account explicitly for a possible change in the length of the cylinder, we now allow it to elongate or contract by axial stretch λ in addition to the torsion (Fig. 5.9). Hence, we compose the two deformations as

$$[F] = \begin{bmatrix} 1 & 0 & 0 \\ 0 & 1 & \gamma r \\ 0 & 0 & 1 \end{bmatrix} \begin{bmatrix} \frac{1}{\sqrt{\lambda}} & 0 & 0 \\ 0 & \frac{1}{\sqrt{\lambda}} & 0 \\ 0 & 0 & \lambda \end{bmatrix} = \begin{bmatrix} \frac{1}{\sqrt{\lambda}} & 0 & 0 \\ 0 & \frac{1}{\sqrt{\lambda}} & \gamma \lambda r \\ 0 & 0 & \lambda \end{bmatrix}. \tag{5.50}$$

Here, $\lambda = h/H$ is the axial stretch, where h and H are the current and reference lengths of the cylinder, respectively.

For the torsion, the upper face of the cylinder is rotated by the angle

$$\Delta\theta = \theta|_{z=h} - \theta|_{z=0}, \tag{5.51}$$

compared to the lower face of the cylinder. It follows that γ, the amount of twist, is $\gamma = \Delta\theta/h = \Delta\theta/(\lambda H)$. Also, comparing the components of F to the general ones in (5.30), we see that here

$$\frac{r}{R} = \frac{1}{\sqrt{\lambda}} = \sqrt{\frac{H}{h}}. \tag{5.52}$$

We can then write F and the right Cauchy-Green deformation tensor $C = F^T F$ in terms of R, $\Delta\theta$, h, and H only, as

$$[F] = \begin{bmatrix} \sqrt{\frac{H}{h}} & 0 & 0 \\ 0 & \sqrt{\frac{H}{h}} & \frac{\Delta\theta}{\sqrt{Hh}} R \\ 0 & 0 & \frac{h}{H} \end{bmatrix}, \quad [C] = \begin{bmatrix} \frac{H}{h} & 0 & 0 \\ 0 & \frac{H}{h} & \frac{\Delta\theta}{h} R \\ 0 & \frac{\Delta\theta}{h} R & \frac{h^2}{H^2} + \frac{\Delta\theta^2}{Hh} R^2 \end{bmatrix}. \tag{5.53}$$

In particular, we find that the first invariant of strain is

$$I_1 = \operatorname{tr} C = 2\frac{H}{h} + \frac{h^2}{H^2} + \frac{\Delta\theta^2}{Hh} R^2. \tag{5.54}$$

We now focus on a cylinder made of a neo-Hookean material with shear modulus μ_0. Then the elastic energy U of the cylinder is obtained by integrating the strain

5.4 Pure Torsion of an Incompressible Cylinder

energy density $W = \frac{\mu_0}{2}(I_1 - 3)$ over the whole volume of the cylinder in the reference configuration:

$$U = \int_0^A \int_0^{2\pi} \int_0^H \frac{\mu_0}{2}(I_1 - 3) dR\, R d\theta\, dZ$$
$$= \mu_0 \pi H \left[\left(2\frac{H}{h} + \frac{h^2}{H^2} - 3 \right) \frac{A^2}{2} + \frac{\Delta\theta^2}{Hh} \frac{A^4}{4} \right], \quad (5.55)$$

where A is the radius of the cylinder in the reference configuration.

The work done by the torque M to twist the cylinder, $W_M = M\Delta\theta$, is constant in this case because we hold $\Delta\theta$ fixed. Therefore, the *total potential energy* E is equal to $E = U - W = U - \text{const}$, and the only unknown is the current length h of the cylinder. Imposing that h minimises the energy, we write that

$$E'(h) = \mu_0 \pi A^2 \left(-\frac{H^2}{h^2} + \frac{h}{H} - \frac{A^2 \Delta\theta^2}{4h^2} \right) = 0. \quad (5.56)$$

We thus obtain the equilibrium length $h^\star$:

$$h^\star = H \sqrt[3]{1 + \frac{A^2 \Delta\theta^2}{4H^2}} > H, \quad (5.57)$$

which immediately shows that the cylinder always elongates upon applying a twist, thus confirming directly the Poynting effect.

At $h = h^\star$, we find that $E'' = 3\mu_0 \pi A^2 / H$, which is always positive, ensuring that the equilibrium is stable.

In practice, it is quite easy to devise an experimental setup showing the effect; see Fig. 5.13.

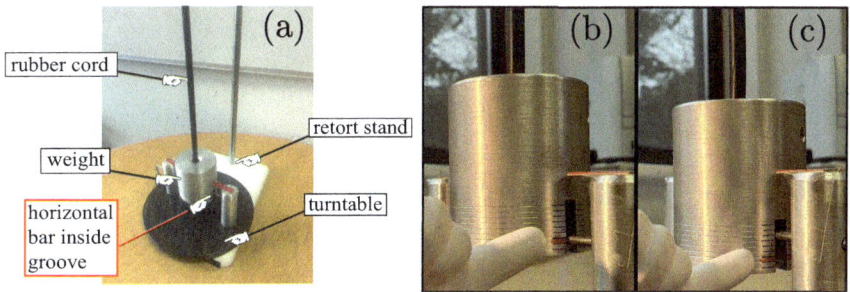

Fig. 5.13 A tabletop demonstration of the Poynting effect in torsion. (**a**) The setup. (**b**) When the cylindrical rubber rod is at rest, the red mark on the weight is at the level of the horizontal metal bar. (**c**) After several twists, the red mark is located lower than the bar, showing that the cord has lengthened. (Reproduced with permission from [2])

5.5 Inflation of a Spherical Membrane

The inflation of thin spherical and tubular membranes is a classical problem of the theory of nonlinear elasticity, often used to assess experimentally the material response of soft polymers (such as rubber and silicone) and of soft biological membranes (such as bladders or veins).

As experience suggests to anyone who has blown air into a party rubber balloon, at first, the balloon is difficult to inflate, and then its radius increases rapidly, with little or no effort. Biological membranes behave quite differently, with an ever-increasing pressure required to inflate them, as shown in Fig. 5.14, which reproduces the experimental pressure-radius curves for the inflation of a rubber balloon and a monkey bladder obtained by Osborne in 1909.

5.5.1 Equilibrium of an Inflated Spherical Membrane

We assume the balloon is a spherical shell of internal radius A and thickness H in the reference configuration, and a and h in the current configuration, respectively. The deformation is described in spherical coordinates, so that a point $\mathbf{X}$ with coordinates (R, Θ, Φ) in the reference configuration is mapped into a point $\mathbf{x}$ with coordinates (r, θ, ϕ) in the current configuration.

When the shell is subjected to an inflation pressure P, it undergoes a spherically symmetric deformation: the spherical shell is transformed into another spherical shell of different radius, so that the deformation is of the form

$$r = \hat{r}(R), \qquad \theta = \Theta, \qquad \phi = \Phi, \tag{5.58}$$

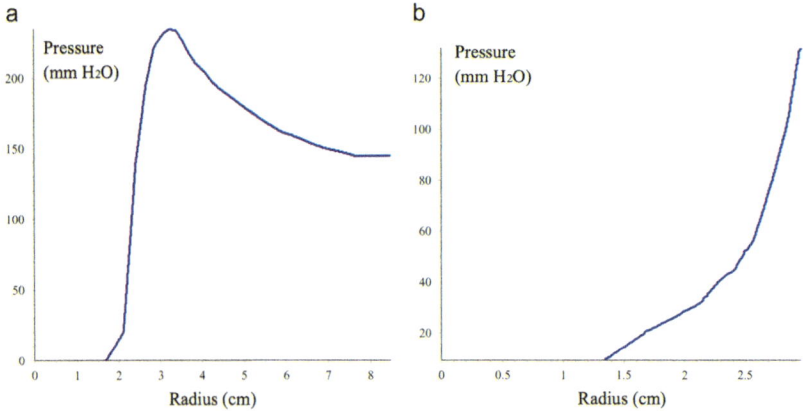

Fig. 5.14 Pressure-radius curves, as recorded by Osborne in 1909. (**a**) Inflation of a rubber balloon. (**b**) Inflation of a monkey bladder

5.5 Inflation of a Spherical Membrane

where $\hat{r}$ is a yet unknown function of R.

The expression for the deformation gradient **F** in spherical coordinates is found in textbooks. In general, it reads

$$[\mathsf{F}] = \begin{bmatrix} \dfrac{\partial r}{\partial R} & \dfrac{1}{R}\dfrac{\partial r}{\partial \Theta} & \dfrac{1}{R\sin\Theta}\dfrac{\partial r}{\partial \Phi} \\ r\dfrac{\partial \theta}{\partial R} & \dfrac{r}{R}\dfrac{\partial \theta}{\partial \Theta} & \dfrac{r}{R\sin\Theta}\dfrac{\partial \theta}{\partial \Phi} \\ r\sin\theta\dfrac{\partial \phi}{\partial R} & \dfrac{r\sin\theta}{R}\dfrac{\partial \phi}{\partial \Theta} & \dfrac{r\sin\theta}{R\sin\Theta}\dfrac{\partial \phi}{\partial \Phi} \end{bmatrix}, \quad (5.59)$$

which here, reduces to

$$[\mathsf{F}] = \begin{bmatrix} \hat{r}' & 0 & 0 \\ 0 & \hat{r}/R & 0 \\ 0 & 0 & \hat{r}/R \end{bmatrix}, \quad (5.60)$$

a diagonal matrix. We identify the diagonal components with the principal stretches in the radial direction, $\lambda_r = \hat{r}'(R)$, and in the polar and azimuthal directions, $\lambda_\theta = \lambda_\phi = \hat{r}(R)/R$.

Because of incompressibility, $\det \mathsf{F} = 1$, and we must solve the following differential equation for $\hat{r}(R)$:

$$\hat{r}'\hat{r}^2 = R^2. \quad (5.61)$$

It can be easily integrated by separation of variables: $\hat{r}^2 d\hat{r} = R^2 dR$, with boundary condition $\hat{r}(A) = a$, where a is the internal radius of the balloon in the current configuration, to obtain

$$\hat{r}(R) = (R^3 + a^3 - A^3)^{1/3}. \quad (5.62)$$

Clearly, the components of **F** in (5.60) are not constants: inflation is a non-homogeneous deformation. From (5.62), we find the stretches in the membrane as

$$\lambda_r = \frac{R^2}{(R^3 + a^3 - A^3)^{2/3}}, \quad \lambda_\theta = \lambda_\phi = \frac{(R^3 + a^3 - A^3)^{1/3}}{R}. \quad (5.63)$$

Note that the deformation (5.58) is completely determined once a is known.

Now we can compute the components of **B** and B^{-1} and substitute into the stress-deformation relationship $\sigma = -p\mathsf{I} + 2W_1\mathsf{B} - 2W_2\mathsf{B}^{-1}$. From textbooks, we find that the equilibrium equation div $\sigma = \mathbf{0}$ in spherical coordinates reads, in general,

$$\frac{\partial \sigma_{rr}}{\partial r} + \frac{2}{r}\sigma_{rr} + \frac{1}{r}\frac{\partial \sigma_{\theta r}}{\partial \theta} + \frac{1}{r\sin\theta}\frac{\partial \sigma_{\phi r}}{\partial \phi} + \frac{\cot\theta}{r}\sigma_{\theta r} - \frac{1}{r}(\sigma_{\theta\theta} + \sigma_{\phi\phi}) = 0,$$

$$\frac{\partial \sigma_{r\theta}}{\partial r} + \frac{2}{r}\sigma_{r\theta} + \frac{1}{r}\sigma_{\theta r} + \frac{1}{r}\frac{\partial \sigma_{\theta\theta}}{\partial \theta} + \frac{1}{r\sin\theta}\frac{\partial \sigma_{\phi\theta}}{\partial \phi} + \frac{\cot\theta}{r}(\sigma_{\theta\theta} - \sigma_{\phi\phi}) = 0,$$

$$\frac{\partial \sigma_{r\phi}}{\partial r} + \frac{2}{r}\sigma_{r\phi} + \frac{1}{r}\sigma_{\phi r} + \frac{1}{r}\frac{\partial \sigma_{\theta\phi}}{\partial \theta} + \frac{1}{r\sin\theta}\frac{\partial \sigma_{\phi\phi}}{\partial \phi} + \frac{\cot\theta}{r}(\sigma_{\theta\phi} + \sigma_{\phi\theta}) = 0. \quad (5.64)$$

Here the only non-zero components of the Cauchy stress are σ_{rr}, $\sigma_{\theta\theta} = \sigma_{\phi\phi}$. So, the second and the third equations above give

$$\frac{\partial p}{\partial \theta} = 0, \qquad \frac{\partial p}{\partial \phi} = 0, \quad (5.65)$$

and we conclude that p and thus the components of $\boldsymbol{\sigma}$, are functions of r only. From this, the first equation of equilibrium now specialises as

$$\frac{d\sigma_{rr}}{dr} + \frac{2}{r}(\sigma_{rr} - \sigma_{\theta\theta}) = 0. \quad (5.66)$$

Because the external surface of the balloon is free of traction, we have $\sigma_{rr}(a + h) = 0$. The internal surface is under the applied pressure: $\sigma_{rr}(a) = -P$. By using these boundary conditions, and upon integration of (5.66), we get

$$P = -2\int_a^{a+h} \frac{\sigma_{rr} - \sigma_{\theta\theta}}{r}\,dr. \quad (5.67)$$

The only unknown in this equation is the current external radius a, so that, at least in principle, this expression provides the pressure-radius relation for the inflated balloon. In practice, however, the integral is quite involved. Here, we consider the special case of a thin-walled balloon, an assumption that strongly simplifies the analysis.

In the so-called *membrane approximation*, the thickness is much smaller than the radius: $H \ll A$, $h \ll a$, so that $R \simeq A$, $r \simeq a$. Then, Equation (5.67) is approximated as

$$P = -2h\frac{\sigma_{rr}(a) - \sigma_{\theta\theta}(a)}{a}, \quad (5.68)$$

where higher-order terms in h are neglected. When the thickness is small, we may also approximate the stretches (5.63) as

$$\lambda_r = \frac{A^2}{a^2} = \frac{1}{\lambda^2}, \qquad \lambda_\theta = \lambda_\phi = \frac{a}{A} = \lambda, \quad (5.69)$$

5.5 Inflation of a Spherical Membrane

where we introduced the notation λ for the *membrane stretch*. Similarly, the stresses are approximated, as

$$\sigma_{rr} = -p + 2W_1\lambda^{-4} - 2W_2\lambda^4, \qquad \sigma_{\theta\theta} = \sigma_{\phi\phi} = -p + 2W_1\lambda^2 - 2W_2\lambda^{-2}, \tag{5.70}$$

evaluated at $r = a$. Note also that W_1 and W_2 are functions of λ, because

$$I_1 = 2\lambda^2 + \lambda^{-4}, \qquad I_2 = 2\lambda^{-2} + \lambda^4. \tag{5.71}$$

Finally, because $a = \lambda A$ according to (5.69), we find that $h = \lambda^{-2} H$, which follows from the conservation of the volume of the material making up the shell $4\pi a^2 h = 4\pi A^2 H$. We then obtain the pressure-stretch relation from (5.68) and (5.70) as

$$P = 4\frac{H}{\lambda^2 A}\left[(\lambda - \lambda^{-5})W_1 - (\lambda^{-3} - \lambda^3)W_2\right]. \tag{5.72}$$

This expression can be put in a more compact form by noticing that, according to (5.71), $W = W(I_1, I_2)$ is in fact a function of λ only. We then introduce the single-valued function $w(\lambda) = W(2\lambda^2 + \lambda^{-4}, 2\lambda^{-2} + \lambda^4)$, and we differentiate it with respect to λ as

$$w' = (4\lambda - 4\lambda^{-5})W_1 + (-4\lambda^{-3} + 4\lambda^3)W_2. \tag{5.73}$$

Then Equation (5.72) takes the compact form

$$P = \frac{Hw'(\lambda)}{A\lambda^2}. \tag{5.74}$$

This is the sought relation between pressure and radius for a thin elastic membrane.

In the case of a Mooney-Rivlin model, with $w(\lambda) = C_1(2\lambda^2 + \lambda^{-4} - 3)/2 + C_2(2\lambda^{-2} + \lambda^4 - 3)/2$, we find that

$$\bar{P} = \left(1 + \frac{C_2}{C_1}\lambda^2\right)\left(\lambda^{-1} - \lambda^{-7}\right), \qquad \text{where} \qquad \bar{P} = \frac{PA}{2C_1 H} \tag{5.75}$$

is a non-dimensional measure of the inflation pressure. Figure 5.15 shows that the corresponding plots can capture the N-shaped curves typical of rubber balloons when C_2/C_1 is small, as well as ever-increasing curves with no maximum typical of biological membranes, when C_2/C_1 is larger than 0.3.

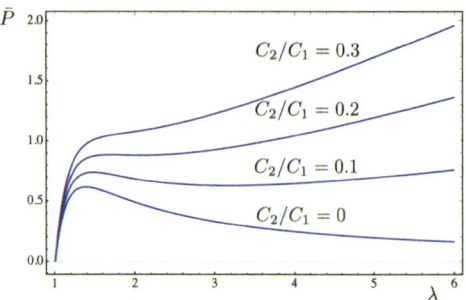

Fig. 5.15 Pressure-radius curves for a spherical shell made of a Mooney-Rivlin material, for different values of the ratio C_2/C_1 between 0 (corresponding to neo-Hookean material) and 0.3

5.5.2 Energy Method for the Inflation of a Spherical Membrane

The principle of minimisation of the total potential energy, already exploited for the Poynting effect in torsion, provides an efficient alternative way to derive the pressure-stretch equation (5.74).

We assume the membrane approximation from the start and describe the membrane deformation through (5.69). Here we use the reduced strain energy function $w(\lambda) = W(2\lambda^2 + \lambda^{-4}, 2\lambda^{-2} + \lambda^4)$ to compute the energy stored inside the membrane as $4\pi A^2 H w(\lambda)$ (because the volume of the elastic membrane is $4\pi A^2 H$.)

Then we look at the work done by the pressure P to inflate the membrane from radius A to radius a. It exerts a force of magnitude $P 4\pi r^2$ on the inner surface of the shell. When the radius increases by dr, the work done is $P 4\pi r^2 dr$. So, going from radius A to radius a requires total work

$$W_P = \int_A^a P 4\pi r^2 dr = \frac{4}{3}\pi P(a^3 - A^3) = \frac{4}{3}\pi P A^3 (\lambda^3 - 1). \tag{5.76}$$

The *total potential energy* of the inflated membrane is, therefore,

$$E = U - W_P = 4\pi A^2 H w(\lambda) - \frac{4}{3}\pi P A^3 \lambda^3 + \text{const}. \tag{5.77}$$

Equilibrium occurs when $E'(\lambda) = 4\pi A^2 (H w' - P A \lambda^2) = 0$. This equation gives the pressure-stretch relation (5.74) directly! It reads

$$P = \frac{H w'(\lambda^\star)}{A \lambda^{\star 2}}, \tag{5.78}$$

where $\lambda^\star$ is the equilibrium stretch.

With this method, we can go further and discuss when a point on the $P-\lambda$ curve corresponds to a stable equilibrium, by looking at where $E''(\lambda) \geq 0$. Hence, we find the condition

$$E''(\lambda) = 4\pi A^2 (Hw'' - 2PA\lambda) \geq 0. \qquad (5.79)$$

Using (5.74), we find that this is equivalent to $\lambda w'' - 2w' \geq 0$, and then a quick check shows that this is the same as

$$P' \geq 0. \qquad (5.80)$$

For example, in Fig. 5.15, for $C_2 = 0$ (lower curve), only the increasing part of the pressure-stretch curve corresponds to stable solutions, and the curve after the maximum represents unstable equilibria.

When $C_2/C_1 = 0.1$ and $C_2/C_1 = 0.2$, we notice that the pressure-stretch curve has two regions with positive slope at the left and right, and a central region with negative slope. This remark helps explain an effect that we commonly encounter when inflating a party balloon: initially the balloon is hard to inflate, corresponding to the high early slope of the curve. Then, suddenly, the balloon becomes easier to inflate, corresponding to the low slope of the curve after the unstable region.

5.6 Electroactive Membranes

The growing need for light, simple, and fast devices capable of converting elastic energy into work (and vice versa) has driven the development of technological applications that couple elasticity with electrostatics. These devices are known as *electroactive polymers*, abbreviated as EAPs.

Their working principle is illustrated in Fig. 5.16. In its reference configuration, an EAP membrane is a thin slab with a thickness H much smaller than its edge length L. The upper and lower surfaces are attached to flexible electrodes, similar to ordinary capacitors, with the difference that the dielectric membrane (typically made of silicone) is highly deformable.

When a voltage V is applied across the electrodes, the membrane tends to thin out. Because of the incompressibility of silicone, this thinning causes the membrane to expand in area. This property is leveraged in advanced applications, such as the development of soft sensors, actuators, and artificial muscles for robotics.

The coupling between electricity and deformation leads to intriguing *multi-physics* properties, which can be complex to describe due to various types of nonlinearities. A significant simplification occurs if we consider *ideal dielectrics*, where the material constants are independent of the EAP state. For these materials, we can analyse the electro-elastic coupling by examining the stresses generated by applied tractions and electric forces, or by following an energetic approach, similar to the approach used for the torsion and inflation problems in the previous sections.

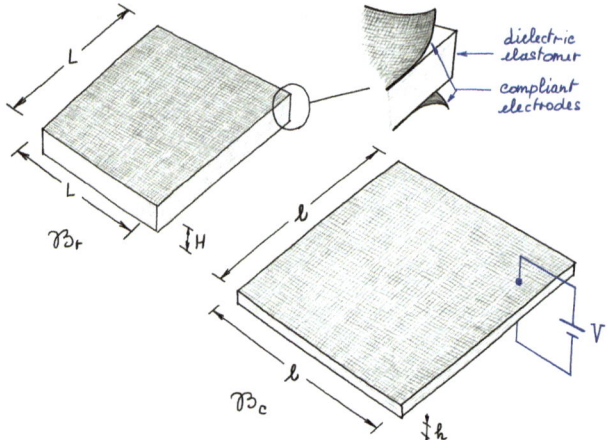

Fig. 5.16 Scheme of an EAP membrane, made up of a dielectric elastomer, sandwiched between two compliant electrodes. Upon applying a voltage V, the membrane is squeezed along the thickness direction and expands in area

5.6.1 Electroelastic Equilibrium in Dielectric Membranes

If we have already studied the topic of Electrostatics (in physics, applied mathematics or engineering courses), we know that the *electric field* is a vector E defined as

$$E = -\operatorname{grad} \varphi, \tag{5.81}$$

where φ is the *electrostatic potential* (Note that to avoid confusion with the basis vectors $\mathbf{E}_i$, we denote the electric field by E.) Also, the volumetric *density of free charges* ϱ_f is linked to the *electric displacement* vector D by the equation

$$\operatorname{div} D = \varrho_f. \tag{5.82}$$

In ideal dielectrics, the electric displacement is related linearly to the electric field through the *dielectric constant* ε as

$$D = \varepsilon E. \tag{5.83}$$

Combining these three equations leads to the Poisson equation for the potential

$$\Delta \varphi = -\frac{\varrho_f}{\varepsilon}. \tag{5.84}$$

5.6 Electroactive Membranes

Consider an infinitesimal volume containing free charges, acted upon by an applied electric field. They experience the following force per unit volume:

$$\mathbf{b}_e = \varrho_f \mathbf{E}, \tag{5.85}$$

the well-known *Coulomb force*. The resultant *electric force* acting upon charges enclosed in a region $\mathcal{R}$ is then

$$\mathbf{f}_e(\mathcal{R}) = \int_{\mathcal{R}} \mathbf{b}_e \, dv. \tag{5.86}$$

Now introduce a tensor $\boldsymbol{\tau}$, with the following components in a local orthonormal basis:

$$\tau_{ij} = \varepsilon \left(E_i E_j - \tfrac{1}{2}(\mathbf{E} \cdot \mathbf{E})\delta_{ij} \right). \tag{5.87}$$

This tensor enjoys the property that its divergence is precisely the Coulomb force:

$$\text{div } \boldsymbol{\tau} = \mathbf{b}_e. \tag{5.88}$$

Checking this identity is a long but straightforward calculation, which we leave as an exercise (it relies on the connections (5.81), (5.84), and (5.85)). This property, together with the divergence theorem, allows us to write the force due to electric fields (5.86) as a surface integral instead of a volume integral

$$\mathbf{f}_e(\mathcal{R}) = \int_{\partial \mathcal{R}} \boldsymbol{\tau}^\mathsf{T} \mathbf{n} \, da, \tag{5.89}$$

making the tensor $\boldsymbol{\tau}$ formally equivalent to the Cauchy stress when it comes to computing resultant forces over a certain region of a dielectric material; see Eq. (3.15) for comparison. The stress due to electrostatic forces, $\boldsymbol{\tau}$, is called the *Maxwell stress*.

When a soft *ideal dielectric* is subjected not only to electric forces but also to mechanical forces, the *total stress*, $\boldsymbol{\sigma}$, may be taken as the sum of the stress due to elastic deformations, $\boldsymbol{\sigma}_{\text{elastic}}$, and the Maxwell stress, $\boldsymbol{\tau}$:

$$\boldsymbol{\sigma} = \boldsymbol{\sigma}_{\text{elastic}} + \boldsymbol{\tau}. \tag{5.90}$$

For example, in an incompressible neo-Hookean ideal dielectric, the total stress is

$$\boldsymbol{\sigma} = -p\mathbf{I} + \mu_0 \mathbf{B} + \boldsymbol{\tau}. \tag{5.91}$$

where μ_0 is the shear modulus in the absence of electricity and p is a Lagrange multiplier due to incompressibility.

We use this expression to study the homogeneous deformation of a soft dielectric plate subjected to an electric field applied along the thickness direction **k**. The counterparts in $\mathcal{B}_c$ to the reference thickness H and edge length L are h and ℓ, respectively. The upper and lower faces are coated with flexible electrodes. The in-plane and through-thickness stretches are

$$\lambda_1 = \lambda_2 = \lambda = \frac{\ell}{L}, \qquad \lambda_3 = \frac{h}{H}, \tag{5.92}$$

respectively. By incompressibility, the volume of the membrane is conserved: $HL^2 = h\ell^2$, so that $\lambda_3 = \lambda^{-2}$. Thus, the left Cauchy-Green deformation tensor of this *equi-biaxial* deformation has components

$$[\mathsf{B}] = \begin{bmatrix} \lambda^2 & 0 & 0 \\ 0 & \lambda^2 & 0 \\ 0 & 0 & \lambda^{-4} \end{bmatrix}, \tag{5.93}$$

in the coordinate system aligned with the edges.

There are no free charges inside the dielectric, so that the Poisson equation (5.84) reduces to the Laplace equation $\Delta\varphi = 0$. Furthermore, as the lateral dimensions are large compared to the thickness, it is reasonable to consider that across the vast majority of the plate's surface, the electric potential does not depend on x_1 and x_2, so that the equation further reduces to $\varphi''(x_3) = 0$, which we integrate twice to $\varphi(x_3) = ax_3 + c$, where a and c are constants. We consider that the electric potential difference between the lower and upper electrodes in $\mathcal{B}_c$ is of value V volts: $\varphi(0) - \varphi(h) = V$, and we deduce that $\varphi(x_3) = -Vx_3/h + c$, and according to (5.81) and (5.92), the electric field is

$$\mathbf{E} = -\frac{V}{h}\mathbf{e}_3 = -\frac{V}{H}\lambda^2 \mathbf{e}_3. \tag{5.94}$$

Because the electric field depends on the deformation itself, it is called a *follower load*, as opposed to *dead loads* which are constant.

We then compute the components of the total stress using (5.90) as

$$[\sigma] = \begin{bmatrix} -p & 0 & 0 \\ 0 & -p & 0 \\ 0 & 0 & -p \end{bmatrix} + \mu_0 \begin{bmatrix} \lambda^2 & 0 & 0 \\ 0 & \lambda^2 & 0 \\ 0 & 0 & \lambda^{-4} \end{bmatrix} + \frac{\varepsilon}{2}\left(\frac{V}{H}\right)^2 \lambda^4 \begin{bmatrix} -1 & 0 & 0 \\ 0 & -1 & 0 \\ 0 & 0 & 1 \end{bmatrix}. \tag{5.95}$$

The equilibrium of the electrical and mechanical forces implies that div $\sigma = \mathbf{0}$, leading here to $p =$ constant. The upper and lower electrodes are free of traction, so that $\sigma_{33} = 0$ there, and in fact everywhere because the components of σ are constants. It follows that

$$p = \mu_0 \lambda^{-4} + (\varepsilon/2)(V/H)^2 \lambda^4, \tag{5.96}$$

5.6 Electroactive Membranes

and

$$\sigma_{11} = \sigma_{22} = \mu_0(\lambda^2 - \lambda^{-4}) - \varepsilon (V/H)^2 \lambda^4. \tag{5.97}$$

We assume that the plate is subjected not only to the voltage V but also to mechanical forces applied on its lateral faces. These lateral faces have surface areas HL in $\mathcal{B}_r$ and $h\ell$ in $\mathcal{B}_c$. Calling s the magnitude of the applied tractions in $\mathcal{B}_r$, we have, at equilibrium,

$$HLs = h\ell\sigma_{11} = h\ell\sigma_{22}, \tag{5.98}$$

which leads to the following relationship between the voltage V, in-plane stretch λ, and applied tractions s:

$$V = H\sqrt{\mu_0/\varepsilon}\sqrt{\lambda^{-2} - \lambda^{-8} - (s/\mu_0)\lambda^{-3}}. \tag{5.99}$$

However, inverting this relationship to find the stretch λ created by the application of voltage and traction is not straightforward because, for some given values of V and s, there may exist multiple values of λ, some corresponding to stable equilibrium states, and others to unstable ones. To decipher the situation, we turn to an energy-based argument.

5.6.2 Energy Method for Electroactive Membranes

To model the coupling of elasticity and electrostatics in the energy-based approach, we postulate that the *electro-elastic energy* of the device can be written as follows:

$$W^*(\mathbf{F}, \mathbf{E}) = W(\mathbf{F}) - \tfrac{1}{2}\mathbf{D} \cdot \mathbf{E}, \tag{5.100}$$

where $W(\mathbf{F})$ is the purely elastic energy, and the second term is the electrostatic energy. The minus sign in the latter may be qualitatively understood by recalling *Thomson's theorem of electrostatics*, stating that when the voltage is prescribed in a system of conductors, the amount of free charge on the electrodes tends to be maximized at equilibrium. Therefore, recalling that in ideal dielectrics $\mathbf{D} = \varepsilon\mathbf{E}$, we see that by *minimizing* $-D^2/(2\varepsilon)$ we effectively *maximize* the magnitude $D = \sqrt{\mathbf{D} \cdot \mathbf{D}}$ which, as we recall, is a measure of the amount of free charge in the system.

For a neo-Hookean dielectric, the electroelastic energy specialises to

$$W^* = \frac{\mu_0}{2}(2\lambda^2 + \lambda^{-4} - 3) - \frac{\varepsilon}{2}\frac{V^2}{H^2}\lambda^4, \tag{5.101}$$

because, according to (5.93), $I_1 = 2\lambda^2 + \lambda^{-4}$. The quantities involved are all homogeneous (the same everywhere), and it is easy to compute the *total potential energy*, E^*, made up of the total internal energy minus the work of edge tractions, as

$$E^* = HL^2 W^* - 2(sHL)(\ell - L) = HL^2[W^* - 2s(\lambda - 1)], \qquad (5.102)$$

because $\ell = \lambda L$. We see that $E^* = E^*(\lambda)$ only.

We can now apply the principle of energy minimisation. First, we write $(E^*)'(\lambda) = 0$:

$$2HL^2\left[\mu_0(\lambda - \lambda^{-5}) - \varepsilon(V^2/H^2)\lambda^3 - s\right] = 0, \qquad (5.103)$$

which gives the voltage-stretch relation (5.99) of the previous section, as expected. Second, we write that $(E^*)'' \geq 0$, or

$$\mu_0(1 + 5\lambda^{-6}) - 3\varepsilon(V^2/H^2)\lambda^2 \geq 0, \qquad (5.104)$$

to find V_c, the critical voltage beyond which the equilibrium becomes unstable, as

$$V_c = H\sqrt{\frac{\mu_0}{\varepsilon}}\sqrt{\frac{\lambda^{-2} + 5\lambda^{-8}}{3}}. \qquad (5.105)$$

We introduce the following non-dimensional measures of voltage and pre-stress,

$$\bar{V} = \frac{V}{H\sqrt{\mu_0/\varepsilon}}, \qquad \bar{s} = \frac{s}{\mu_0}, \qquad (5.106)$$

to plot the curves

$$\bar{V} = \sqrt{\lambda^{-4} - \lambda^{-8} - \bar{s}\lambda^{-3}}, \qquad \bar{V}_c = \sqrt{\frac{\lambda^{-2} + 5\lambda^{-8}}{3}}. \qquad (5.107)$$

Figure 5.17 displays the loading voltage-stretch $\bar{V} - \lambda$ curves for different levels of pre-stress $\bar{s} = 0.0, 1.0, 2.0, 3.0$, and the curve (dotted line) giving the location of $\bar{V}_c$, located at the maximum of each curve. The figure is consistent with several fundamental effects observed in dielectric thin plates in the lab; see Fig. 5.17.

When the membrane is loaded by voltage only (no pre-tension, $\bar{s} = 0$), we see a steep curve at first, as an increase in the voltage leads to a modest increase in the stretch. Then a maximum is reached, and, theoretically, the voltage cannot be increased further. In practice, we can of course increase the voltage beyond that maximum, but then the behaviour of the plate is no longer described by the homogeneous deformations we just studied.

5.7 The Biot Instability

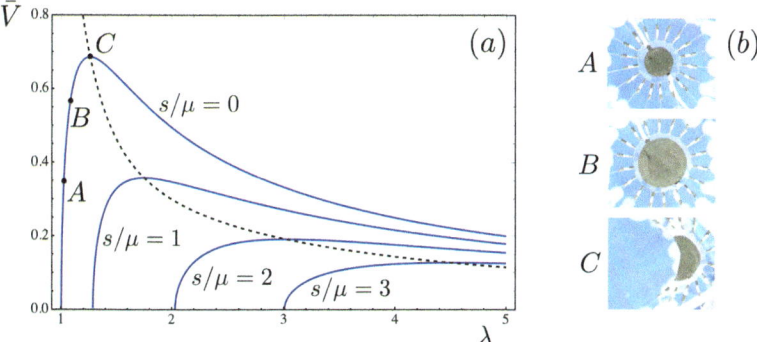

Fig. 5.17 (**a**) Voltage-stretch curves for increasing magnitude of pre-stress $s/\mu = 0, 1, 2, 3$, showing that the limit stress for pull-in can be delayed by pre-tensioning the dielectric thin plate. Dotted curve: plot of the maxima, representing the critical voltages beyond which equilibrium is unstable on the remainder of each voltage-stretch curve. (**b**) Experimental evidence of voltage-driven expansion in area for $s = 0$, until the stretching limit is reached and the membrane snaps ("pull-in" instability) in C

Next, we look at the curves when there is a pre-tension, $\bar{s} > 0$. Then the maximum is lowered and not as sharp, and can even disappear, so that greater gains in stretch can be achieved. This observation led to the discovery that pre-stretch strongly enhances the performances of EAPs.

5.7 The Biot Instability

Our daily experience of natural and engineering phenomena suggests that elastic bodies, when compressed sufficiently, can exhibit a sudden qualitative and quantitative change in their deformation path. This phenomenon goes by the name of *elastic instability*. It can be appreciated, for example, by compressing a straight ruler along its length: above a certain level of compressive force, known as the Euler critical load [1744], it suddenly bends instead of remaining straight. This aspect is crucial in the design, for example, of slender trusses, where elastic instability can anticipate structural failure and have catastrophic consequences.

Elastic instabilities also characterise the behaviour of soft solids, as illustrated by the surface corrugations of the gel in the shear box of Fig. 2.8: at a critical value of the compression along the shortest diagonal, the homogeneous deformation of the gel becomes unstable, and wrinkles appear on its surface.

The study of elastic stability in nonlinear elasticity is a wide and complex topic, where analytical treatments are rare. One way to make progress is to model the apparition of wrinkles on the boundary of a compressed solid. Mathematically, we consider a known large amplitude solution, such as those presented in this chapter, and extend it with a small-amplitude perturbation in the shape of wrinkles.

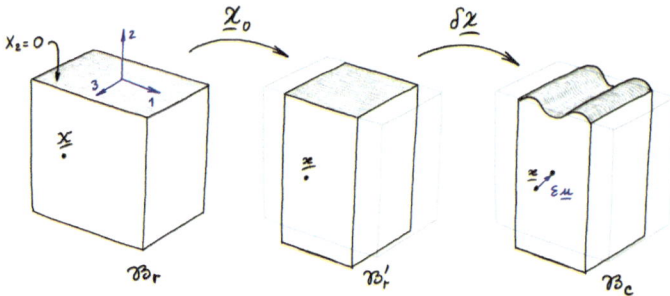

Fig. 5.18 A semi-infinite soft solid is subjected to a homogeneous deformation $\mathbf{x} = \chi_0(\mathbf{X})$ bringing a point at $\mathbf{X} = (X_1, X_2, X_3)$ in the reference configuration $\mathcal{B}_r$ to a point at $\mathbf{x} = (x_1, x_2, x_3) = (\lambda X_1, X_2/\lambda, X_3)$ in the so-called intermediate configuration $\mathcal{B}'_r = \chi_0(\mathcal{B}_r)$. A further small-amplitude perturbation with additional displacement $\epsilon \mathbf{u}$ makes the current configuration $\mathcal{B}_c$ unstable when $\lambda = \lambda_c$

In this section, we treat a simple problem: how much can an elastic half-space made of incompressible neo-Hookean material be compressed before wrinkles appear on its surface, as schematised by Fig. 5.18. This effect is commonly known as "Biot instability", after the Belgian Applied Mathematician Maurice Biot [1905–1985], who solved it in 1963 (slightly earlier, two Irish Mathematical Physicists, Mike Hayes [1936–2017] and Jim Flavin [1936–2012], had independently and separately arrived at similar results).

Consider that the neo-Hookean half space occupies the region $X_2 \le 0$ in the reference configuration $\mathcal{B}_r$. Denote by $(\mathbf{E}_1, \mathbf{E}_2, \mathbf{E}_3)$ the orthonormal basis, and assume that the material deforms homogeneously with constant principal stretches occurring along the axes directions. The principal stretch along $\mathbf{E}_1$ is $\lambda_1 = \lambda < 1$, as the material is confined in such a way that it is prevented from expanding along $\mathbf{E}_3$, so that $\lambda_3 = 1$ (this is called a plane strain deformation). The boundary $X_2 = 0$ is free of traction.

The solid deforms homogeneously until, at the *critical stretch* $\lambda = \lambda_c$, wrinkles appear on the surface. The position of a point initially at (X_1, X_2, X_3) in $\mathcal{B}_r$ is thus at $(\lambda_1 X_1 + \epsilon u_1, \lambda_2 X_2 + \epsilon u_2, \lambda_3 X_3 + \epsilon u_3)$ in $\mathcal{B}_c$, where $\epsilon \ll 1$ (because, at least initially, the amplitude of wrinkles is small) and $\mathbf{u}$ is the "incremental" mechanical displacement. At this juncture, we remark that in principle, we should find it easier to write the boundary condition of a traction-free surface in $\mathcal{B}_r$, where it is flat, than in $\mathcal{B}_c$, where it is wrinkled. We thus adopt the Lagrangian description and write the displacement as a function of $\mathbf{X}$. Moreover, we look for wrinkles that vary in the (X_1, X_2) plane only (and not in the X_3 direction); see Fig. 5.18, and take $\mathbf{u} = [u_1(X_1, X_2), u_2(X_1, X_2), 0]^T$. It follows that the deformation gradient has components

$$[\mathsf{F}] = \begin{bmatrix} \lambda + \epsilon u_{1,1} & \epsilon u_{1,2} & 0 \\ \epsilon u_{2,1} & \lambda_2 + \epsilon u_{2,2} & 0 \\ 0 & 0 & 1 \end{bmatrix}, \qquad (5.108)$$

5.7 The Biot Instability

where the comma denotes partial differentiation with respect to the Lagrangian coordinates (for example, $u_{2,1} = \partial u_2/\partial X_1$).

From now on, we expand expressions up to first order in ϵ. Thus, the incompressibility constraint $\det \mathsf{F} = 1$ reads $\lambda \lambda_2 + \epsilon(\lambda_2 u_{1,1} + \lambda u_{2,2}) = 1$. At order zero in ϵ, it gives $\lambda \lambda_2 = 1$, so that $\lambda_2 = \lambda^{-1}$. Then, at order one, we have

$$\lambda^{-1} u_{1,1} + \lambda u_{2,2} = 0, \tag{5.109}$$

which we can solve by introducing a "stream function" ψ, a neat trick that you might have encountered in Fluid Mechanics. Here $\psi = \psi(X_1, X_2)$ is such that

$$u_1 = -\lambda \frac{\partial \psi}{\partial X_2}, \qquad u_2 = \lambda^{-1} \frac{\partial \psi}{\partial X_1}, \tag{5.110}$$

and is defined by integrating its derivatives above. Instead of two unknown functions u_1, u_2, we now have only one, ψ, and the so-called *incremental incompressibility condition* (5.109) is automatically satisfied (check!).

According to (4.58) and (3.29), the nominal stress in the neo-Hookean material is of the form $\mathsf{S} = -p\mathsf{F}^{-1} + \mu \mathsf{F}^\mathsf{T}$, where μ is the shear modulus and p is the Lagrange multiplier due to incompressibility. To be consistent, we expand it in ϵ as $p = p_0 + \epsilon p_1$, where p_0 and p_1 are functions of X_1, X_2.

We then compute F and its inverse (up to the first order in ϵ) in terms of ψ as

$$[\mathsf{F}] = \begin{bmatrix} \lambda(1 - \epsilon \psi_{,12}) & -\epsilon \lambda \psi_{,22} & 0 \\ \epsilon \lambda^{-1} \psi_{,11} & \lambda^{-1}(1 + \epsilon \psi_{,12}) & 0 \\ 0 & 0 & 1 \end{bmatrix},$$

$$[\mathsf{F}^{-1}] = \begin{bmatrix} \lambda^{-1}(1 + \epsilon \psi_{,12}) & \epsilon \lambda \psi_{,22} & 0 \\ -\epsilon \lambda^{-1} \psi_{,11} & \lambda(1 - \epsilon \psi_{,12}) & 0 \\ 0 & 0 & 1 \end{bmatrix}, \tag{5.111}$$

which we use to find the components of S, as

$$S_{11} = -p_0 \lambda^{-1} + \mu \lambda - \epsilon[p_1 \lambda^{-1} + (p_0 \lambda^{-1} + \mu \lambda)\psi_{,12}],$$
$$S_{12} = \epsilon(-p_0 \lambda \psi_{,22} + \mu \lambda^{-1} \psi_{,11}),$$
$$S_{21} = \epsilon(p_0 \lambda^{-1} \psi_{,11} - \mu \lambda \psi_{,22}),$$
$$S_{22} = -p_0 \lambda + \mu \lambda^{-1} + \epsilon[-p_1 \lambda + (p_0 \lambda + \mu \lambda^{-1})\psi_{,12}],$$
$$S_{13} = S_{23} = S_{31} = S_{32} = 0, \quad S_{33} = -p_0 + \mu - \epsilon p_1. \tag{5.112}$$

Writing the equations of equilibrium $\text{Div}\,\mathsf{S} = \mathbf{0}$ at the zero-th order in ϵ shows that $p_{0,1} = p_{0,2} = 0$, i.e. p_0 is a constant.

The boundary condition of a traction-free surface is $S^T E_2 = 0$ at $X_2 = 0$, or

$$S_{21} = 0, \quad S_{22} = 0, \quad S_{23} = 0, \quad \text{at } X_2 = 0. \tag{5.113}$$

Using (5.112), we see that the third condition is trivially satisfied. At the zero-th order, we find that the first equation is also satisfied and that, according to the second equation, $p_0 = \mu \lambda^{-2}$. We now have

$$\begin{aligned} S_{11} &= \mu(\lambda - \lambda^{-3}) - \epsilon[p_1 \lambda^{-1} + \mu(\lambda + \lambda^{-3})\psi_{,12}], \\ S_{12} &= \epsilon \mu \lambda^{-1}(\psi_{,11} - \psi_{,22}), \\ S_{21} &= \epsilon \mu(\lambda^{-3}\psi_{,11} - \lambda \psi_{,22}), \\ S_{22} &= \epsilon(-p_1 \lambda + 2\mu \lambda^{-1}\psi_{,12}). \end{aligned} \tag{5.114}$$

The first two equations of equilibrium are $S_{11,1} + S_{21,2} = 0$ and $S_{12,1} + S_{22,2} = 0$ (the third one is trivial), and they yield the following expressions for the derivatives of p_1:

$$\frac{\partial p_1}{\partial X_1} = -\mu \lambda^2(\psi_{,112} + \psi_{,222}), \quad \frac{\partial p_1}{\partial X_2} = \mu \lambda^{-2}(\psi_{,111} + \psi_{,122}). \tag{5.115}$$

Finally, writing that the crossed second-order partial derivatives $\partial^2 p_1/\partial X_1 \partial X_2$ and $\partial^2 p_1/\partial X_2 \partial X_1$ are the same leads to a partial differential equation for ψ

$$\lambda^4 \psi_{,2222} + (1 + \lambda^4)\psi_{,1122} + \psi_{,1111} = 0. \tag{5.116}$$

To solve it, we model the wrinkles as sinusoidal patterns in $\mathcal{B}_c$. Because there is a one-to-one correspondence between the coordinates in the current and reference configurations, and because the mechanical displacement is derived from ψ, it makes sense to look for solutions in the following separable form:

$$\psi(X_1, X_2) = f(X_2) \sin(k X_1), \tag{5.117}$$

where $k > 0$ is a constant and f an unknown function of X_2. Following substitution into (5.116), we end up with a fourth-order ordinary differential equation with constant coefficients for f:

$$k^4 \lambda^4 f'''' + (1 + \lambda^4)k^2 f'' + f = 0. \tag{5.118}$$

We look for exponential solutions $f(X_2) = e^{rX_2}$ and find the four roots of the characteristic equation as $r = \pm k$ and $r = \pm k \lambda^{-2}$. We discard the positive roots, because they would lead to solutions exponentially growing with depth, which

5.7 The Biot Instability

would be unphysical (and also contradict the assumption of small amplitude). We conclude that

$$\psi(X_1, X_2) = \left(ce^{-kX_2} + de^{-k\lambda^{-2}X_2}\right)\sin(kX_1), \tag{5.119}$$

where c, d are constants.

Then, integrating either equation in (5.115) gives

$$p_1(X_1, X_2) = \mu k^2 d(1 - \lambda^{-4})e^{k\lambda^{-2}X_2}\cos(kX_1). \tag{5.120}$$

We can now write the two traction-free boundary conditions (5.113)$_{1,2}$ at the first-order in ϵ, which leads to a homogeneous system of two equations for the constants c, d:

$$(1 + \lambda^4)c + 2d = 0, \qquad 2\lambda^2 c + (1 + \lambda^4)d = 0. \tag{5.121}$$

Writing that the determinant of this system is zero to ensure non-trivial solutions yields the following polynomial equation for λ:

$$\lambda^8 + 2\lambda^4 - 4\lambda^2 + 1 = 0, \tag{5.122}$$

with the following positive roots: 0.5436890127 and 1. We discard $\lambda = 1$, because according to (5.114)$_1$, (5.119), (5.120), and (5.121) written when $\lambda = 1$, we would then have $S_{11} = 0$ and a spontaneous instability without compression, which is unphysical.

We conclude that

$$\lambda_c \simeq 0.54, \tag{5.123}$$

which means that, in principle, the semi-infinite neo-Hookean body can be compressed by about 46% before its surface wrinkles. Remarkably, the result is independent of the shear modulus and of the constant k, which is linked to the wavelength of the wrinkles. As shown by Biot, the result carries over to the whole class of Mooney-Rivlin materials (4.59), independently of the material constants C_1, C_2.

In practice, it proves impossible to observe wrinkles on the surface of a compressed homogeneous block of soft matter, because creases appear before λ_c is reached. Creases are a nonlinear phenomenon, for which the first-order perturbation approach is no longer adequate. However, sinusoidal wrinkles do develop on the surface of a soft block coated with a stiffer film, and the method followed in this subsection can be applied to accommodate this setup. These two phenomena, creases on a homogeneous block and wrinkles on a coated block, are readily observed in the shear box experiments with jelly of Chap. 2, see Fig. 2.8.

References

1. M. Destrade, M.D. Gilchrist, J.G. Murphy, B. Rashid, G. Saccomandi. Extreme softness of brain matter in simple shear. *International Journal of Non-Linear Mechanics*, 75 (2015) 54–58.
2. G. Zurlo, J. Blackwell, N. Colgan, M. Destrade. The Poynting effect. *American Journal of Physics*, 88 (2020) 1036–1040.

Glossary

Biot Instability An instability phenomenon where a semi-infinite elastic material under compression exhibits wrinkles on its surface (Chap. 5, Sect. 5.7).

Cauchy's Theorem A foundational result in continuum mechanics stating that the stress vector on any plane within a material can be uniquely determined by the Cauchy stress tensor, which is symmetric and describes the internal forces (Chap. 3, Sect. 3.2).

Constitutive Equation A mathematical expression relating stress and strain for a material, capturing its mechanical behaviour. In nonlinear elasticity, these equations often depend on material-specific stored energy functions (Chap. 4).

Continuum An idealised representation of matter, where materials are assumed to be continuously distributed without discrete particles. This assumption is fundamental to the mathematical formulation of continuum mechanics (Chap. 2, Sect. 2.1).

Deformation Gradient A second-order tensor that describes the mapping of infinitesimal line elements from the reference configuration to the current configuration of a deformable body. It provides a comprehensive measure of local deformations, including changes in length, area, and volume (Chap. 2, Sect. 2.2).

Electroelasticity The study of materials that deform under combined mechanical and electrical forces. This is relevant for electroactive polymers and possibly to biological tissues with electrical stimuli (Chap. 5, Sect. 5.6).

Eulerian Description A perspective in continuum mechanics where field quantities are described in terms of the spatial coordinates of the current configuration (Chap. 2, Sect. 2.1).

Homogeneous Deformations A type of deformation where every material point undergoes the same transformation (the deformation gradient is independent of the position). Examples include simple elongation, pure dilation, and simple shear, all of which have practical applications in understanding material behaviour (Chap. 2, Sect. 2.9).

Hyperelasticity A material model in which the stress-strain relationship is derived from a stored energy function. This framework is commonly applied to rubber-like and biological soft materials that undergo large deformations (Chap. 4, Sect. 4.3).

Invariants of a Symmetric Tensor Coefficients of the characteristic equation of the tensor. They remain unchanged under coordinate transformations (Chap. 1, Sect. 1.7).

Isotropy A property of materials where the material response is identical in all directions, implying invariance under rotations of the reference configuration (Chap. 4, Sect. 4.3).

Material Symmetry The invariance of a material's constitutive behaviour under a set of transformations, such as rotations or reflections, defined by its symmetry group (Chap. 4, Sect. 4.3).

Nanson's Formula A vector relation connecting areas in the reference and current configurations through the deformation gradient and its determinant (Chap. 2, Sect. 2.4).

Lagrangian Description A perspective in continuum mechanics where field quantities are described in terms of the material coordinates of the reference configuration (Chap. 2, Sect. 2.1).

Polar Decomposition Theorem A mathematical theorem stating that any deformation gradient can be uniquely decomposed into a product of a proper orthogonal tensor (rotation) and a symmetric positive-definite stretch tensor (Chap. 2, Sect. 2.6).

Poynting Effect A nonlinear elasticity effect observed when a twisted material exhibits an axial elongation during pure torsion (Chap. 5, Sect. 5.4.2).

Principal Axes of Strain and Stress The axes along the eigenvectors of the stretch and stress tensors, with the associated eigenvalues representing the principal stretches and stresses (Chap. 3, Sect. 3.4).

Strain Energy Density Function A scalar function that relates the deformation of a material to its stored elastic energy per unit volume, commonly used in hyperelastic material modelling (Chap. 4, Sect. 4.7).

Stress Tensor A tensor representing the internal forces within a material (Chap. 3, Sect. 3.3).

Tensor Algebra The mathematical framework used to describe and manipulate quantities like scalars, vectors, and second-order tensors. In the context of continuum mechanics, it is essential for analysing stress, strain, and deformation gradients (Chap. 1, Sect. 1.5).

Index

B
Biological soft tissues, 1–3, 21
Biot instability, 65, 97–101
Boundary conditions, 49–50, 75, 88, 98, 100, 101

C
Cauchy's theorem, 40, 41–44
Constitutive equations, 51–64
Curve fitting, 64, 69–71, 74, 82

D
Deformation gradient, 19–20, 24, 25–27, 30, 32, 34, 50–52, 54, 60, 78, 87, 98
Differential operators, 14–15, 20
Divergence theorem, 36–37, 42, 43, 49, 50, 93

E
Electroactive membranes, 65, 91–97
Equilibrium equations, 43, 44, 49, 87

F
Fung model, 76, 80

G
Gent model, 64, 70, 71, 76

H
Homogeneous and inhomogeneous deformations, 65–66
Homogeneous deformations, 29–36, 65–66, 94, 96–98
Hydrostatic pressure, 46, 47, 50

I
Inflation of a balloon, 86–91
Isotropic hyperelasticity, 55–57, 62, 65, 66

L
Linear algebra, 5

M
Material symmetry, 55–57
Mooney-Rivlin model, 63, 69–71, 74–76, 80–82, 89, 90, 101

N
Neo-Hookean model, 62, 63, 68–70, 74–76, 84, 90, 93, 95, 98, 99, 101
Nominal stress tensor, 47–49
Nonlinear elasticity, 1, 2, 5, 10, 20, 49, 65–101

O
Objectivity, 53–57
Ogden model, 64

P
Polar decomposition theorem, 54
Principal stresses, 46
Principle of work, 51–53

R
Rubber-like material, 1, 3, 4

S
Shear stress, 45, 47, 72–75, 77
Simple shear, 23–25, 31–33, 65, 72, 74–77, 81, 82
Stored energy functions, 62–64
Strain measurement, 29

Stress-strain curve, 4
Stress tensor, 39, 41, 43–46, 48, 51, 78
Surface traction, 40–43, 50, 52, 65

T
Torsion of a cylinder, 66

U
Uni-axial stress, 47
Uni-axial tension, 47, 65, 67

W
Work, 20, 39, 50–53, 60–62, 83, 85, 90, 91, 96

The manufacturer's authorised representative in the EU is Springer Nature Customer Service Centre GmbH, Europaplatz 3, 69115 Heidelberg, Germany. If you have any concerns regarding our products, please contact ProductSafety@springernature.com

Printed and bound by CPI Group (UK) Ltd, Croydon, CR0 4YY
23/02/2026
02058934-0001